D1045697

THE CONCEPT OF
ENERGY
SIMPLY EXPLAINED

(formerly titled: The Story of Energy)

by Morton Mott-Smith, Ph.D.

Illustrated by Emil Kosa, Jr.

Dover Publications, Inc., New York

Copyright © 1964 by Dover Publications, Inc.

All rights reserved under Pan American and International Copyright Conventions.

This Dover edition, first published in 1964, is an unabridged and revised version of the work first published by D. Appleton-Century Company in 1934 under the former title: *The Story of Energy*.

The publisher wishes to thank the directors of the Oberlin College Library for making a copy of this book available.

International Standard Book Number: 0-486-21071-5

Library of Congress Catalog Card Number: 63-19496

Manufactured in the United States of America

Dover Publications, Inc.
New York 14, N. Y.
180 Varick Street

OUR MECHANICAL CIVILIZATION

OUR present civilization differs from all previous ones in the possession of mechanical power. This has in the last century brought about more rapid and more profound changes in human life and relations than has occurred in all the previous centuries combined. It has produced an enormous and rapid increase in population and in population density. Rapid transportation and instant communications have spread this civilization to all parts of the earth, so that it is no longer subject to attack from without, but only to disintegration from within.

This disintegration would be brought about at once by a failure in the supply of power. Nearly all of the billion horsepower we now employ is obtained from the combustion of fuel, and most of that from the burning of fossil fuel, coal and oil. These resources are fast approaching exhaustion. Water, wind, wave, tidal, and sun power, and the internal heat of the earth, have been used to some extent, but even if fully developed would scarcely meet our present needs, let alone any future increase. Men are looking to science for new sources of energy. Hope for the future is based on atomic energy.

Yet this marvelous age of power and of fast living is little more than a century old. It began with Watt's invention of the condensing steam-engine in 1765; but many decades elapsed before the use of power became general. Fulton's *Claremont* steamed up the Hudson in 1806. The first railways were built in 1830. The Otto gas engine appeared in 1876, the Parsons' turbine in 1884, the Diesel engine in 1894. Trolley cars appeared in the nineties. The automobile, the airship, and the airplane belong to the twentieth century. The rate of progress until now has been continually accelerated.

Our present huge population and our present mode of life are made possible only by the mass production of the necessities of life with the aid of mechanical power. Should this fail, many of us would perish at once. We have lost the primitive arts, and even much of the semi-civilized arts, and cannot at once return to them. We are like domesticated animals—no longer able to live in the wild. Humanity could survive only by a drastic reduction in its numbers. Indeed, we may in future have to reduce our numbers (as Wells foresees), and practise the strictest economy of energy. We shall have to live within our income, and cease spending our capital. If, as seems probable, humanity has yet a billion years before it, this whole boasted age of progress, which seems already on the wane, will be looked back upon as but a moment of youthful extravagance, a period of irreparable destruction rather than of construction. Our great cities, like the pyramids, will be looked upon as monuments of stupidity, and the people of the future will wonder how those of this age could waste so much energy in so useless a manner.*

In this book we intend to trace the story of man's conquest of energy, to describe the scientific discoveries that made it possible, and the chief ways in which energy is applied to useful purposes. The greatly accelerated pace of the last decades of the century of power was made possible by a revolution in scientific ideas that occurred in the middle of the nineteenth century. This was brought about by the discovery of the conservation of energy, and the establishment of the mechanical theory of heat. They produced also a profound change in our conception of the universe. Accordingly they will occupy a large place in our story.

Grateful acknowledgements are due to Mr. James T. Barkelew, who painstakingly read the manuscript, corrected errors, and made many valuable suggestions; to Mr. Emil Kosa, the artist who did the freehand drawings; to Mr. Russell Porter for the illustration of a modern steam-engine; and to Mr. Carl Nagashima, who lettered the drawings.

M. M. S.

---

* Atomic energy or some other form still not "harnessed" may save us yet.

CONTENTS

ILLUSTRATIONS

CHAPTER I

THE QUEST FOR POWER

EVERY animal, every plant, every living thing is perpetually seeking energy. Food, air, and sunshine are the chief sources. To be sure, the energy required by living things is very small compared with the amounts involved in the titanic operations of inorganic nature—such as wind, wave, volcano, or the radiation of a star—but it is none the less essential. Much of the food we eat has no other value than its energy content. Sugar and most fats build no tissue. They are fuel only.

Man is the only animal that consciously seeks and utilizes energy other than that of his own body. First beasts of burden supplemented his muscles, then water and windmills, and the sail boat. But the wind does not always blow, and running water is not everywhere available. Man needed a source of power that was independent of time and place.

A beginning in this direction was made by Hero of Alexandria in the first century A.D. He found that air, expanded by heating, could be made to produce a mechanical force, and he used it to actuate many sorts of mechanisms, such as his famous fountain, or to open the doors of a temple mysteriously when a fire was built on the altar, and to close them when the fire was extinguished. He was the first outstanding mechanician—one who studies to convert a simple push, pull, or rotation into other desired motions. His most famous invention for the production of mechanical power was the steam turbine illustrated in Figure 1. Water was contained in the covered and sealed bowl, and heated by a fire beneath. Steam passed up through the hollow supports and through hollow bearings into the globe. From there it issued tangentially through the two nozzles shown, and

FIG. 1. HERO'S STEAM TURBINE

by its reaction caused the globe to revolve in the opposite direction.

The reaction principle is now applied to lawn sprinklers, pinwheels, toy rockets, and to other small appliances. It was used by De Laval in 1882 in a small steam turbine applied to a cream separator. But reaction alone is very inefficient. All of our great modern turbines are actuated by the impact or the pressure of steam, or both, in addition to reaction. But of this later.

All of Hero's contrivances were buried soon after by the onset of the Dark Ages. Fifteen centuries later they were dug up and repeated with variations. Giovanni della Porta in 1601 operated a Hero's fountain with steam instead of air. Other sources of light mechanical power were discovered. The draught of a chimney or the convection current above a flame was used to turn a light vaned wheel; such was Leonardo da Vinci's invention for turning a roasting spit. A jet of steam impinging on a vaned wheel was used by Branca in 1629, thus anticipating the impulse turbine. But all of these contrivances were mouse-power toys. No one then dreamed of developing a mechanical power, other than that of the wind and water-mills already existing, comparable in magnitude with a man's muscles; and because of the cheapness of human and animal labor, and the simplicity of life, there was indeed no urgent need of it.

The invention of mechanical clocks and watches in the thirteenth century stimulated a great interest in complicated mechanisms. Great skill and ingenuity were developed. Mechanical marvels, especially automata which imitated the actions and sounds of men or animals, like Vaucanson's duck, or played musical instruments, became very popular. The impression grew that anything could be accomplished by mechanism. In philosophy, Descartes declared that an animal was simply a complicated machine; and Lamettrie extended the idea to man himself.[1]

An outstanding figure of this mechanistic age was Leonardo da Vinci. His note-books are filled with all sorts of mechanical devices. And he did not confine himself to small and light contrivances, but conceived huge machines, mostly engines of war, operated by droves of men or animals. Only a mechanical power of large magnitude was wanting to make his inventions practical.

Early efforts to produce mechanical power turned largely, however, to perpetual motion. To get something for nothing, seems to have been man's perennial desire. These contrivances began to appear in the twelfth century, and thereafter multiplied rapidly, despite the fact that Stevinus as early as 1595, and Galileo a little later, denied their possibility, and showed that the basic laws of mechanics are incompatible with them.[2] The quest for perpetual motion began to wane only when real mechanical power became cheap toward the middle of the nineteenth century. Nevertheless it still goes on to some extent, and doubtless

[1] Lamettrie, *L'homme-machine* (Leyden, 1748).
[2] Mach, *Die Mechanik* (Leipzig, 1883). Morton Mott-Smith, *Principles of Mechanics Simply Explained* (New York, Dover, 1963) pp. 39, 60.

at this very moment some one somewhere is working on such a contraption.

A little consideration will show that any engine does precisely what the perpetual motionist seeks; it gives something for nothing. The prime object of mechanical power is to save human labor. The perpetual motionist expects to save the labor of supplying the fuel. He does not expect to save the labor of constructing the engine. The machines he has so far built have been small and cheap, but, while no perpetual motionist has ever given the power of his machine or any means by which it could be calculated, it is not to be supposed that a machine to deliver a hundred thousand horse-power would be the size of a pin-wheel. Large forces, however produced, require stout pieces to transmit and handle them. Since no one has ever built a perpetual motion machine that would work, we cannot say how its cost would compare with that of a steam- or gas-engine of the same power. But, at any rate, the first cost is not saved.

A certain amount of labor is required to build a steam- or gas-engine. A certain further amount is required to tend, repair, and provide it with fuel. But during its lifetime, the engine gives back far more work than was expended upon it in all of these ways. If it did not, it would not be built. This excess is given absolutely gratis. No one has paid for it one cent of money or one ounce of labor. Yet we can get money for it. We can buy human labor with it. We get something for nothing.

At the present time the world is using billions of horse-power. Since a mechanical horse-power according to Watt is equivalent to the power of seven men, this is equivalent to the power of several times the world's entire population. Possibly half of this population is not working, and the other half works only a third of the time. Our engines not only save labor, they produce labor far beyond the capacity of the whole human race.

Moreover, in many respects mechanical labor is much superior to that of men or animals. It is more rapid, more regular, more exact; it never tires, it can go on day and night; the machine eats nothing when idle, and when worn out can be thrown on the scrap heap—and no one has to take care of its old age. Indeed, the metals of which it is composed can be melted down and used over again. But, most important of all, because of the concentration of great power in a small space and weight, the engine can do what men and beasts, whatever their numbers, could never do. Consider, for example, an automobile equipped with a sixty-horse-power engine. The power of the engine is equal to that of 420 men. An hour's work of these men at one dollar an hour would cost $420. The engine in that time will consume four or five gallons of gasoline costing perhaps over a dollar. But while the 420 men can do the same quantity of work that the engine can, they could not send the machine and themselves flying at fifty or sixty miles per hour. Still less could they lift an aëroplane loaded with their weight into the air. Leonardo da Vinci's flying machine might have been a success, if he could have found a fifty-horse-power man to operate it. Now, unless a perpetual motion machine were vastly cheaper to build than a gasoline engine of the same power, all that it could save in the above instance would be the dollar and change paid for the fuel.

Finally, we have fuelless motors in our water, wind, tide, wave, and sun motors. With these, as with a perpetual motion, the first cost and the maintenance are the whole cost. They are, to be sure, not transportable. But we have ways of transmitting and storing the energy they produce. While the first cost of a large hydro-electric plant is indeed enormous, an efficient plant of the sort can produce electricity at a cost of one half mill per kilowatt-hour. Now a kilowatt is equivalent to the power of ten men. At one dollar an hour for human labor, such a plant does the work of 20,000 men at the cost of one man. Perpetual motion is therefore superfluous. We already have all it pretended to offer, and more.

The Middle Ages did discover a real source of power, however, in the explosion of gunpowder. While this substance had long been known in China, its use as a propellant for musket and cannon balls was a distinctly European invention. A gun is essentially an engine, in that it converts the expansive force of an explosion into mechanical motion, but it is hardly adapted to the driving of machinery. Nevertheless, Huygens in the seventeenth century proposed to utilize it for that very purpose. His plan was to explode a small charge of powder at the bottom of a long vertical cannon, and by this means to elevate a weight which fitted the gun like a piston. The weight in descending was to pull up another weight, or to actuate directly any desired machinery. Curiously enough, a century and a half later, this idea was actually carried out in what were called free-piston engines, only instead of gunpowder, gas was used as the explosive.

Other forms of explosive engine had been experimented with as early as 1800. Lycopodium powder, the substance used for stage fire, and other finely powdered combustibles were tried. Little success was achieved, however, until illuminating gas came into use, and even then progress was slow because of the high price of this fuel. In the meantime, the power of steam had attracted attention, and soon achieved so great a success that all other types of engine were for a while abandoned.

CHAPTER II

THE EARLY STEAM-ENGINE

THERE IS an old story to the effect that James Watt as a boy once sat in his mother's kitchen holding a spoon over the spout of a kettle, and noted how the steam thus held back caused the lid of the kettle to lift. In this way, it is recounted, he discovered the power of steam and when he grew up, invented the steam-engine. It's a pretty story, but unfortunately it's not true. The power of steam was known and steam-engines were in use long before Watt was born in 1736. In 1601 Giovanni della Porta, in a treatise on pneumatics, called attention to the force of steam when confined, and—what is more important—to the fact that when steam is condensed in a closed chamber a vacuum is created that can suck up water. Both of these principles were applied to a pumping engine invented by Thomas Savery in 1698. This was nothing more than a force pump in which the down stroke was produced by the pressure of steam on top of the water, and the up or suction stroke by sousing the pump chamber with cold water, thereby condensing the contained steam and creating a vacuum which sucked up water from the well below. A small pump called the pulsometer, operating on this principle, is used today. Savery's pump, having no moving parts, was not a true engine in the sense that it could produce power for general purposes. There was indeed no great need of such power at the time. But there was pressing need of power pumps. The coal mines in England were fast filling with water, and the whole industry was threatened with extinction. The power pump came just in time to save the situation. Appropriately Savery called his pump "The Miner's Friend."

The first true engine of commercial importance was the New-

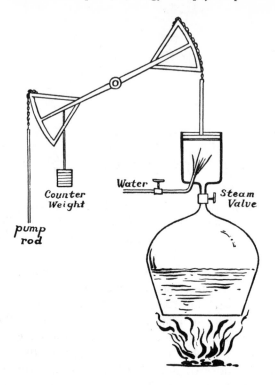

FIG. 2. THE NEWCOMEN ATMOSPHERIC ENGINE

comen atmospheric engine of 1705. This was a piston and cylinder engine, and thus, having moving parts, could have been applied to any purpose. But it was only used for pumping. It
consisted of a large vertical cylinder set directly on top of the
boiler, as shown in Figure 2. The piston was pulled up by a
counterweight attached to the other end of a walking beam.
This permitted steam to flow from the boiler into the cylinder,
for the steam being at atmospheric pressure—whence the name
of the engine—did no pushing. The piston had to be pulled up.
When the piston arrived at the top of its stroke, the steam was
shut off, and cold water was sprayed into the cylinder. This
condensed the steam, producing a vacuum. The cylinder being
open at the top, the pressure of the air then pushed the piston
down, raised the counterweight and pulled up the pump rod.

This was the power stroke. The water was then drained off from the bottom of the cylinder, and the operations repeated.

In these old engines, the valves were all opened and closed by hand. The story is told that in 1713 a lad named Humphrey Potter, who had been hired for this purpose, was much annoyed by friends who used to come and play games just outside the engine room. So young Potter rigged up an arrangement of ropes and rods, by which the motion of the engine itself was made to open and close the valves at the proper moments, while he went out to play with his friends. But his ingenuity cost him his job, and also those of all future valve boys, for the device, improved in various ways, has become the automatic valve gear now used on all engines.

In the year 1763, James Watt, then a young instrument maker at the University of Glasgow, was given a model of the New-comen engine to repair. Having put it in order, he was astonished by the amount of steam the little engine consumed. He suspected that a good part of the hot steam on entering the cold cylinder, chilled by the previous water spray, condensed, and thus contracting its volume allowed more steam to enter. But it was not easy to test this hypothesis. It was easy enough to determine the *weight* of steam consumed in a given time, by the measuring the feed-water meanwhile supplied to the boiler—the water level being kept constant. Every pound of water consumed became a "pound of steam" and passed through the engine. The *volume* of steam consumed at each revolution, supposing no condensation, would be that of the cylinder. But if condensation occurred, the volume of steam coming from the boiler would be greater than that of the cylinder. It became necessary then to determine the volume a pound of steam would have if no condensation occurred. The necessary condition for this, Watt found, was that the vessel containing the steam should be at all times at the same temperature as the steam.

No such measurements had ever been made. But at that time, Joseph Black, the discoverer of the latent heat of steam, who knew more about steam than any other man then living, was at the university, and helped Watt to make them. The latter found that one volume of water at atmospheric pressure produced 1800 volumes of steam. This result led to the rule of thumb long used by engineers, that a cubic inch of water becomes a cubic foot of steam (1728 volumes). Neither of these figures is correct, the more accurate value being 1600 volumes, and this is true only at atmospheric pressure. For higher pressures the steam volume is less, and for lower pressures it is larger.

With the aid of his measurements, Watt found that the little Newcomen engine consumed no less than eight cylinderfuls of steam at every revolution. In short, seven-eighths of the entering steam at once condensed, and being shrunk to the insignificant volume of water, only one-eighth remained as steam to create, by its subsequent condensation, the vacuum that worked the engine. Or, to put it another way, seven-eighths of the entering steam was employed in merely reheating the cylinder walls to the temperature of the steam, in order that the remaining eighth might survive as steam to work the engine.

The remedy for all this, Watt said, was to keep the cylinder always as hot as the entering steam. At first he thought this could be done by making the cylinder of nonconducting materials, such as wood; but this proved impractical. He returned to the metal cylinder, but surrounded it with a steam jacket, fed with steam directly from the boiler, and further wrapped the whole with heat insulating materials. The benefit of the steam jacket, however, is problematical; for to keep the cylinder hot steam must condense in the jacket to replace the heat lost by radiation, and it is a question whether this steam might not have been better employed in doing work in the engine.

Watt's final and effective remedy was to condense the steam in a separate chamber, where a vacuum was constantly maintained. This chamber, when put in communication with the engine cylinder at the proper moments, sucked the steam out of the latter, thus producing in the cylinder the vacuum required to pull down the piston. And this it did more expeditiously and more effectively than when the vacuum was produced by the spray. But the greatest benefit of the condenser was that it separated the hot and the cold parts of the engine. The cylinder was always hot, the condenser always cold. Thus was avoided the alternate heating and cooling of the same vessel that was responsible for the wastefulness of the Newcomen engine. The principle of condensing the steam elsewhere than in the cylinder was an epoch-making improvement, and more than anything else made possible the modern efficient steam-engine.

Watt's next improvement was to raise the pressure of the steam above that of the atmosphere, so that the steam could push as well as the vacuum pull. Every stroke thus became a power stroke. The *capacity* of the engine was increased, that is, more power was obtained for the same size of cylinder. Watt showed further that the steam pressure might be so far raised that the condenser could be dispensed with altogether, and the engine exhaust directly to the atmosphere. He made no use of this idea, however, and stuck always to the condensing engine.

All of these improvements and some others were embodied in Watt's first patent taken out in 1769. This is probably the most momentous and comprehensive patent ever issued, for it covered the whole production of power by steam. In 1782, he patented two further improvements. The first of these was the double-acting cylinder, in which the steam is admitted alternately on either side of the piston while being exhausted from the other side, thus again doubling the capacity of the engine. All steam-engines are now double-acting. The second improvement was the expansive working of the steam, and this, next to the separate condenser, was his most important improvement.

Up to this time, steam had always been admitted to the cylinder during the full length of the stroke. Consequently, when the exhaust valve opened at the end of the stroke, the cylinder was still filled with steam at the full boiler pressure, and this steam blew out into the condenser with great force and violence, the pressure dropping abruptly from that of the boiler to that of the condenser. It occurred to Watt that this violent expulsion of the steam meant that it still contained a considerable amount of power that might have been used in pushing the piston, if the expansion consequent upon the pressure drop had taken place in the cylinder instead of in the exhaust pipe, where it merely made a rumpus and choked up the condenser. He saw that the way to obtain this power was to admit steam for only a portion of the stroke, and then, with all valves closed, allow the imprisoned steam to expand with gradually diminishing pressure for the rest of the stroke. When the exhaust valve then opened, the pressure would have sunk to nearly that of the condenser, and all the power of the steam would have been utilized, except for a small amount that it is necessary to leave in order to make the steam exit promptly. This arrangement would of course reduce the capacity of the engine, for the mean pressure of the steam during the stroke would be less than the boiler pressure. If the admission of steam is cut off at one-fourth stroke, the work obtained from a given cylinder is reduced about one-half. But only a quarter of a cylinderful of steam is used. Hence the economy is doubled.

All engines now use expansive working of the steam, except direct-acting pumps and locomotives when starting up. The latter are equipped with a *variable* valve gear which enables the moment of closing the admission valve, the cut-off as it is called, to be varied. In starting, the engine driver sets his cut-off late in the stroke so as to obtain the large power necessary for starting. The exhaust being directed up the smokestack to increase the draught, we then hear the loud *choo-choo* and see the immense volumes of steam that are spouted forth, both evidences of great waste. As the train picks up speed, the engine driver gradually advances his cut-off to earlier points in the stroke; the noise and steam clouds diminish, and the engine is running more economically.

Watt used only three or four expansions. That is, he expanded his steam to three or four times the volume it had at cut-off. It is not uncommon to-day to expand the steam two hundred times.

With these several improvements, Watt's engines had an economy around eight times that of the Newcomen. They also had greater capacity, speed, and smoothness of running; and they were applicable to all power purposes. The Newcomen was never used for anything but pumping, and it is doubtful if it could very well have been used for anything else. It was ponderously slow, on account of the time required to fill the cylinder with steam, and, at the end of the down stroke, to drain off the water. In appearance Watt's engine resembled the Newcomen, for he still used a large vertical cylinder and the overhead walking beam to transmit the motion to a fly-wheel at the other end.

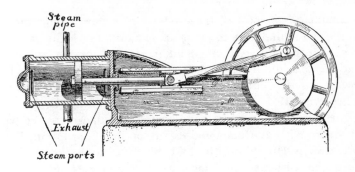

FIG. 3. A SIMPLE MODERN STEAM-ENGINE

The neat-looking compact modern engine, shown in Figure 3, differs only mechanically from Watt's engine. Its action is exactly the same and its economy is not much better. Steam is alternately admitted and exhausted through the rectangular openings or *ports* seen at each end of the cylinder. By means of a valve gear on the back side of the cylinder, and hence not visible in the drawing, each port is put into communication with the steam or with the exhaust pipe, or kept closed altogether, at the proper times. There are many types of valve gear, but being merely mechanisms, we will not discuss them. The motion is communicated through the piston rod, cross-head, connecting rod, and crank, to the fly-wheel. Speed and power are regulated either by a throttle valve which controls the steam supply, or by an arrangement by which the moment of cut-off can be varied.

When the piston is at either end of the stroke, the piston rod, connecting rod, and crank, are all in line, and no push or pull on the piston can move the fly-wheel. The engine is then on a *dead center*. To start the engine, the fly-wheel must be turned by hand or otherwise until the engine is off the dead center. To make a steam-engine always self-starting, at least two cylinders must be employed with their cranks at an angle to each other. Then when either is on a dead center, the other will be in a position to push.

In 1768, the firm of Boulton and Watt was formed, and by virtue of Watt's basic patents, the duration of which was extended by Act of Parliament, this firm enjoyed for thirty years a complete monopoly of the production of power. This proved in the end a hindrance to progress. The firm, as usually happens, became conservative, crushed all rivals and opposed all innovations that did not emanate from itself. High pressure had been proposed, but Boulton and Watt stuck to *seven* pounds. They even tried to have a law passed prohibiting the use of a higher pressure, on the ground that it was dangerous. Perhaps at the time it was. Compound expansion had been invented and patented by Hornblower in 1781, but since he could not build an engine without conflicting with Watt's patents, he was completely paralyzed. Yet high pressure and compounding were precisely the improvements next in order, and destined ultimately to improve the Watt engine, as much as he had improved the Newcomen. But they had to await the end of the monopoly of Boulton and Watt in 1802.

High pressure, which here means 30 to 50 pounds, was introduced by Trevithick around 1800. His single-cylinder non-condensing engine was long popular because of its simplicity and cheapness. The immediate object of high pressure was to increase the capacity of the engine and hence to lessen its cost. There are other advantages, but they were not at the time appreciated, and were not to any extent obtained. We shall discuss them later.

The compound expansion engine was revived by Woolf in 1804. In this type of engine the steam is partially expanded in a small high-pressure cylinder, and is then led to a larger low-pressure cylinder where the expansion is completed. This double expansion engine must be distinguished from the simple two-cylinder engine, like that of the usual locomotive, in which the cylinders are of the same size and steam is led to each directly from the boiler, and which has only the object of making the engine self-starting.

A two-cylinder engine, having four instead of only two power strokes for each revolution, produces a more even turning of the engine, and a lighter fly-wheel may be used. The advantage of additional power strokes, however, was not obtained by Woolf because he used two vertical cylinders working on the *same* end of the walking beam. Also in a modern tandem engine, in which two cylinders are placed end to end and work on the same piston rod, no additional power strokes are obtained. Nevertheless, even in these cases a more even turning of the engine is obtained.

Suppose, for example, that we have a single cylinder in which the cut-off occurs at one-ninth stroke. The volume of the steam at the end of the stroke will be nine times what it was at cut-off, and the pressure will also be reduced to approximately one-ninth of the boiler pressure. It is easy to see that there will be a powerful but short-lived push at the beginning of the stroke, and that the pressure thereafter drops rapidly. This would give a jerky motion to the engine and require a large fly-wheel. Suppose now we divide this ninefold expansion into two *stages*. Each cylinder will have to give only a threefold expansion, or in general, the square root of the total number. This means that in each cylinder cut-off will occur at one-third stroke. The maximum push will last three times as long as before, and will only be reduced at the end of the stroke to one-third, instead of one-ninth, of its original value. The maximum push being thus prolonged, and the pressure fluctuations diminished, the running of the engine is more even.

But this advantage is not gained without expense. Two cylinders cost more than one, and furthermore, since the low-pressure cylinder must in the end accommodate the whole of the ninefold expanded steam, it must be just as large as though the whole expansion had taken place in this cylinder alone. The expansion starts, in the low-pressure cylinder, at one-third stroke, with the steam already expanded three times; but instead of this expansion having been produced by moving the piston from one-ninth to one-third stroke, it has been produced in another cylinder. Few men at that time could see any worth-while advantage in the extra expense of the high-pressure cylinder, and so the Woolf engine met with little success. For half a century the single-cylinder high-pressure engine continued to be the favorite.

The manner in which compound engines were finally introduced is very curious. It had become necessary to increase the power of certain plants using the old Watt engines. Instead of junking the old engines and buying new and larger ones, it occurred in 1845 to an engineer named M'Naught, simply to add another cylinder to work on the other end of the walking beam. The steam pressure was then raised, so that the steam, after partial expansion in the new cylinder, could still work the old cylinder in the usual way. The expected increase in power was of course obtained, but also a totally unexpected improvement in economy.

Yet this magical combination of high pressure with compound expansion, which practical men thus discovered accidentally while seeking something else, had been available since the days of Watt; and furthermore its advantage had been pointed out on theoretical grounds twenty-one years previously by one Sadi Carnot. But of him we shall speak later.

CHAPTER III

THE ENGINE INDICATOR

JAMES WATT was no mere mechanician. His mechanical abilities were of the highest order, but he was also a *steam engineer*, the first to merit that title. He realized that a steam engineer should know something about steam, and it was to such knowledge indeed that his greatest achievements were due. But it is not enough to know merely the general properties of steam. It is highly desirable to know just what the steam is doing in an engine cylinder, when the engine is running. And so Watt devised a little instrument, the *indicator*, which tells us that very thing, which enables us, so to speak, to X-ray the cylinder and see what is going on inside. The instrument is applicable not only to steam, but to any sort of piston and cylinder engine, such as gas, oil, and hot-air engines. It has become indispensable to all those who study, test, design, or build such engines. It is hard to say which was ultimately the more valuable, Watt's engine improvements or his indicator. The former gave us at once a vastly better engine, but the latter has helped us to build still better ones. It has enabled us to test new ideas and theories. It has taken the guess-work out of engine design. It has made possible scientific engine research.

Essentially Watt's indicator, which is diagrammed in Figure 4, is a miniature recording steam gauge. It consists of a small vertical cylinder containing a piston, the latter normally held down by a stiff spiral spring. The top of the cylinder is in communication with the outside air, so that atmospheric pressure prevails above the piston. When steam, or other gas under pressure, is admitted under the piston, the latter is pushed up against the opposition of the spring, a distance proportional to the pressure. If the steam or gas is at a pressure less than that of the atmosphere, the piston is sucked down, the spring distended an amount proportional to the *degree of vacuum.* The instrument therefore measures pressures up from and down from the atmospheric.

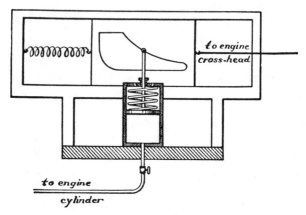

FIG. 4. WATT'S ENGINE INDICATOR

When the indicator is put in communication with one end of an engine cylinder, its little piston moves up and down in accordance with the variations of pressure in the cylinder. In order to make a graphic record of these variations, and to bring them into relation with the motion of the engine, Watt placed a vertical card behind the indicator, and caused this card to slide back and forth edgewise and horizontally in unison with the strokes of the engine. The motion was taken from the crosshead of the engine, but communicated to the card through a reducing mechanism, which reduced it in a known proportion. A pencil, moved by the piston rod of the indicator, when pressed against the card, traced out a shoe-shaped diagram, like that of Figure 5, which showed the pressure in the cylinder for every position of the engine piston, both on its outward and on its return stroke. Modern indicators are more elaborate than the one illustrated, but the principle is the same.

Figure 5 is a typical "indicator card" taken from a non-condensing engine. The diagram is traced in the direction of the arrow, the upper lines during the forward stroke, which is to the right, the lower ones during the return stroke. The horizontal "atmospheric line" is drawn by the indicator when the steam from the engine is shut off, and the card alone is moving. Vertical distances above and below this line are proportional to pressures above and below the atmosphere respectively. The lowermost horizontal line is drawn in afterwards by hand, and is set below the atmospheric line a distance corresponding to the atmospheric pressure at the time, as determined by a barometer reading. It is therefore the axis of zero pressure. Heights measured up from this line give *absolute pressures*. Heights measured up or down from the atmospheric line give *gauge pressures*. It follows that a constant pressure line on this diagram is horizontal.

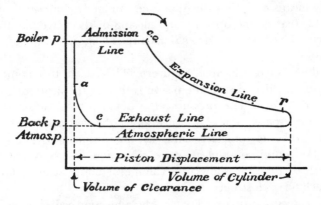

FIG. 5. INDICATOR CARD OF A STEAM-ENGINE

Horizontal distances measured to the right from the back of the shoe are proportional to the travel of the engine piston from the beginning of its forward stroke. Since the volume of a cylinder is proportional to its length, these distances are proportional to the volumes uncovered or *displaced* by the piston. The volume of the cylinder itself is always a little greater than the total piston displacement, because the piston must not hit the cylinder ends. A little space is therefore left, though in steam-engines it is made as small as possible. The volume of this space, including that of the steam port up to the valve, is called the *clearance*. It must be determined by measurements on the engine itself. The left-hand vertical axis of the diagram is then put in by hand, and is set back from the shoe a distance proportional to the clearance volume. It is the axis of zero volume, and distances measured from it to the right, are total volumes of the steam in the cylinder. It follows that a constant volume line on this diagram is vertical.

It is obvious that Watt's indicator diagram is a *plot*. The pressures of the steam are plotted vertically, and the corresponding volumes horizontally. Every point on the diagram represents a particular combination of pressure and volume, the former being given by the height of the point above the base line, the latter by its distance from the left-hand axis. It is a pressure-volume diagram, or as we call it for short, a *pv* plot.

By means of such a diagram, one can study the *events of the stroke,* that is, the opening and closing of the valves, and their effects. These events are—*admission, cut-off, release,* and *compression.*

Admission occurs at *a,* Figure 5. That is, at this point, the admission valve opens. The piston is then at the beginning of its forward stroke. The pressure at *a* is that of the steam in the clearance space. It rises at once along the vertical back of the shoe to boiler pressure, and, as the piston moves forward, the horizontal constant pressure *admission line* is traced. The length of this line shows the volume of boiler steam admitted at each forward stroke.

At *c.o.* cut-off occurs. That is, the admission valve closes. The steam being now in a completely closed chamber, as the piston advances and increases the volume, the pressure drops, and the down-sloping *expansion line* is traced.

At *r* release occurs. That is, the exhaust valve opens, and the pressure drops at once to the exhaust or *back pressure.* The piston now returns, pushing the steam out through the exhaust port, and the *exhaust line* is traced. The back pressure is always a little higher than the atmospheric, because the steam must be *pushed* out. In a good engine, with ample ports and exhaust pipe, the exhaust and atmospheric lines will be almost identical. The difference has been exaggerated in the drawing for clearness. If the engine is condensing, the exhaust line will lie below the atmospheric line, and slightly above the condenser pressure.

At *c,* somewhat before the end of the return stroke, *compression* occurs. The exhaust valve closes, and the steam caught in the clearance space is compressed along the line *ca*. This steam is called the *cushion steam,* because it helps to arrest the backward motion of the piston. It is very important in high-speed engines. Similar cushion steam at the other end of the cylinder helps to arrest the forward motion.

If the events of the stroke are not properly timed, the valves too slow, the steam passages too small, the steam unduly wet, or anything else is faulty in the design or adjustment of the engine, it will show on the indicator card. The engineer reads these cards as a doctor reads his cardiograms. They show the heart-beats of the engine, and from them the engineer can diagnose the trouble.

Another valuable use of the indicator is to measure the power of an engine. Work is force times distance, as pounds times feet or *foot-pounds*. Hence the total push of the steam on the piston, times the distance the latter moves, gives the work done. Now the total push on the piston is equal to the steam pressure (pounds per square foot) times the area of the piston (square feet). Thus to find the work, we multiply together—pressure, area, and distance. But the area of the piston times the distance it moves, is the volume it displaces by its motion. It is the *change* in volume that the steam in the cylinder undergoes. Hence we may say that the work done by the steam is equal to its pressure times its change in volume. Calling these three quantities W, p, and dv (difference in volume), respectively, we may put this statement in the form $W = p \times dv$, which is a convenient abbreviation. Obviously the formula applies only so long as the pressure is constant.

Now let us consider the work done by the steam during admission. The pressure is constant, and is represented on the indicator card by the height p of the admission line. The change in volume is given by the length of this line. The product of these two gives the work. But this product is obviously also equal to the area of the shaded rectangle in Figure 6. This area is proportional then to the work done by the steam during admission.

To find the work done during expansion is less simple, because the pressure is not now constant. But we may get around the difficulty in the following way. Let P, Figure 6, be any point on the expansion line. The pressure at this point is given by its vertical height above the base line. Now let us, keeping the pressure constant, move the piston forward a very short distance, displacing a small volume, *dv*. The work done will be $p \times dv$ as before, and will be represented by the area of the first of the vertical strips shown in the diagram. Now, with the piston stationary, let us lower the pressure to the value it has on the curve, and then, keeping it constant at this lower value, move the piston forward another short distance. The work done will be represented by the area of the second strip on the diagram, and so on. We may thus descend the whole expansion line by a flight of stairs that follows its course very closely, and the whole work then done will be represented by the total area of a pile of strips, which area closely approximates that under the curve. The smaller and more numerous we make the steps, the closer are the approximations. If we make the steps infinitesimal in size and infinite in number, they follow the curve exactly and the total area obtained will be exactly equal to the area under the curve, and between the verticals dropped from the starting and finishing points. This area then represents the work done

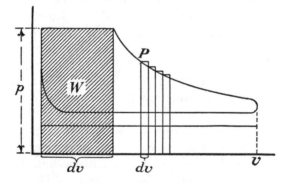

FIG. 6. WORK DONE DURING ADMISSION AND EXPANSION

during expansion. Evidently this kind of summation can be applied to a curve of any form. It is called *integration*.

During the forward stroke the steam does work on the piston. During the return stroke the piston does work on the steam, pushing it out through the exhaust port and compressing a part of it. This latter work may be supplied by the energy stored in a fly-wheel, or by the steam pushing on the other side of the piston in a double-acting engine. The work is shown by the shaded areas in Figure 7.

The net work that we obtain from the four operations is evidently the sum of the first two, diminished by the sum of the last two. Obviously this is equal to the area bounded by the shoe-shaped diagram itself. And this conclusion is true, whatever the precise shape of the diagram.

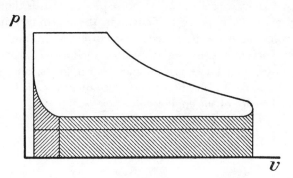

FIG. 7. WORK DURING EXHAUST AND COMPRESSION

The work that an engine does can therefore be found by measuring the areas of its indicator cards. This can be done in several ways, but the most convenient is to use a planimeter. This is a little instrument, which, by running a pointer around the outline of the diagram, indicates the area on a dial. The scales of the card being known, this area can be at once translated into foot-pounds. If the engine is double-acting, we must attach an indicator also to the other end of the cylinder, and measure the work done at that end. If there are several cylinders, we must attach indicators to each. Adding all together, we get the total work done during one revolution of the engine.

Power is the *rate* of doing work. To work faster, one must work harder. That requires a more powerful man, or a more powerful engine. Power is therefore measured in foot-pounds per minute. The unit generally employed, the horse-power, as established by Watt, is a rate of 33,000 foot-pounds per minute, or 550 per second. Having determined the work done by the engines in one revolution, we multiply by the number of revolutions per minute, and dividing by 33,000 get the *indicated* horsepower.

Another way to measure the power of an engine, is to apply a brake to the rim of the fly-wheel or to a special drum, and measure the tangential force exerted on the brake band. Multiplying this by the linear distance traveled by a point on the rim in one minute, gives again the foot-pounds per minute, and dividing by 33,000, we get the *brake* horse-power. This is the power which the engine is ready to deliver to the machinery it is to run. It is less than the indicated horse-power by the power consumed by friction and by the accessories, such as feed, condenser or oil-pumps, etc. These are all mechanical losses, and so the ratio of the brake to the indicated horse-power is called the mechanical efficiency of the engine.

CHAPTER IV

THE MOTIVE POWER OF HEAT— CARNOT, 1824

WATT'S great improvements in the steam engine were mostly due, as we have seen, to a superior knowledge of steam. He applied science to engineering, and improved the *economy* of engines. His immediate successors, however, did not follow his example. Their efforts were directed mainly toward increasing the *capacity* of engines, with a view to reducing their cost. This was the natural aim of practical men. Any further improvement in economy was purely accidental, and came as a complete surprise. Yet to get the same power with less fuel is surely also a matter of practical importance, because it lessens the cost of power. But while capacity is a *mechanical* problem, economy is a *thermal* problem. While the former might yield to the method of fumble and blunder, as practised by practical men, the latter required the orderly theoretical investigation of a trained scientist, acquainted with the phenomena of heat. It was not a task for the mechanically-minded engineer, but for a thermally-minded thinker.

Such a man was Sadi Carnot (granduncle of the Sadi Carnot who became President of France in 1887 and was assassinated in 1894), a young artillery officer, who used to get frequent and prolonged furloughs in order to pursue his studies in mathematics and physics. During the course of his brief life (1796–1832), he wrote but one book, but a book that has been pronounced the most marvelous hundred pages in the literature of science. This book, in which he laid down for all time the basic principles that govern the efficiency of engines, he wrote in 1824, when he was but twenty-eight years of age.

We speak of steam power, as though steam itself were a source of power. But steam is only a carrier of energy. The real source of its power is the fire under the boiler. Similarly, gas, oil, and hot-air engines, all derive their power from the combustion of fuel. All are, as Carnot dubbed them, *heat-engines*. The problem before him therefore was to discover the source of the motive power of fire, and the most efficient means of obtaining it. That this was the problem, he indicated immediately in the title of his book—*Reflections on the Motive Power of Fire and on Machines Fitted to Develop that Power**—and in its opening words, which are:

> Every one knows that heat can produce motion. That it possesses vast motive power no one can doubt, in these days when the steam-engine is everywhere so well known.
>
> To heat also are due the vast movements which take place on the earth. It causes the agitations of the atmosphere, the ascension of clouds, the fall of rain and of meteors, the currents of water which channel the surface of the globe, and of which man has thus far employed but a small portion. Even earthquakes and volcanic eruptions are the result of heat.
>
> From this immense reservoir we may draw the moving force necessary for our purposes. Nature, in providing us with combustibles on all sides, has given us the power to produce, at all times and in all places, heat and the impelling power which is the result of it. To develop this power, to appropriate it to our uses, is the object of heat-engines.
>
> The study of these engines is of the greatest interest, their importance is enormous, their use is continually increasing, and they seem destined to produce a great revolution in the civilized world. . . .
>
> Steam navigation brings nearer together the most distant nations. It tends to unite the nations of the earth as inhabitants of one country. In fact, to lessen the time, the fatigues, the uncertainties, and the dangers of travel—is not this the same as greatly to shorten distances?

These prophetic words, we should remember, were written in 1824, prior to the first public railways, and when the Atlantic had been traversed but once by a steamship, the *Savannah* in 1819! Further on Carnot says:

> Notwithstanding the work of all kinds done by steam-engines, notwithstanding the satisfactory condition to which they have been brought

---

* *Reflections on the Motive Power of Fire by Sadi Carnot and Other Papers on the Second Law of Thermodynamics by E. Clapeyron and R. Clausius*, edited by E. Mendoza, was published by Dover Publications in 1960.

to-day, their theory is very little understood, and the attempts to improve them are still directed almost by chance.

And then:

In order to consider in the most general way the principle of the production of motion by heat, it must be considered independently of any mechanism or any particular agent. It is necessary to establish principles applicable not only to steam-engines but to all imaginable heat-engines, whatever the working substance and whatever the method by which it is operated. . . . We have long sought, and are seeking to-day, to ascertain whether there are in existence agents preferable to the vapor of water for developing the motive power of heat; whether atmospheric air, for example, would not present in this respect great advantages. We propose now to submit these questions to a deliberate examination.

It is obvious from these observations that the efficiency of a steam-engine is not properly measured by the amount of coal, or even of steam, consumed in proportion to the power developed, but by the *quantity of heat* employed. This measure alone is applicable to all heat-engines.

At the time that Carnot wrote, heat was supposed to be an imponderable fluid, to which Lavoisier had given the name *caloric*. This fluid could enter or leave bodies and thereby raise or lower their temperatures. Being a material substance, it was uncreatable and indestructible. All the phenomena of heat had therefore to be explained by redistributions of this fluid, without alteration in its total quantity.

Carnot was not fully satified with the caloric theory, but since no other was officially recognized, he had to adopt it. His basic principles, however, are quite independent of any hypothesis regarding the nature of heat.

We must pause for a moment in our discussion of Carnot to remind the reader of a few elementary facts concerning heat and the properties of steam. When two bodies of different temperatures are placed in contact, the temperature of the one falls while that of the other rises, until both assume the same temperature. We explain this by supposing that something, heat or caloric, passes from one to the other, and that the amount of heating or cooling depends upon the quantity of heat that passes. We therefore measure heat by its heating or cooling effect. The quantity of heat required to raise the temperature of a gram of water one degree on the centigrade scale, is the French or metric unit of heat, the *calorie*. To raise a gram of water two degrees, or to raise two grams one degree, requires two calories of heat, and so on. In short, the heat required is proportional both to the rise in temperature and to the mass of the water that absorbs it. To measure a quantity of heat then, we have only to see how much it will raise the temperature of a measured quantity of water to which it is communicated. The process is called *calorimetry*, and the apparatus a water *calorimeter*. There are of course other methods—the steam, the ice calorimeter, etc.—but all must be reduced to the water calorimeter.

The French calorie is a very small unit, because the gram is only a thimbleful of water. For practical purposes, the kilogram-calorie is much used. This is equal to a thousand small calories. We shall spell it with a capital C—Calorie.

In the English system, the heat unit employed is the quantity of heat required to raise a pound of water (about a pint) one degree Fahrenheit. It is equal to 252 gram-calories or about one-fourth of a large Calorie. It is called the British Thermal Unit, often abbreviated to B.T.U. We shall call it the *therm*.*

The quantity of heat required to raise the temperature of a unit mass of *any* substance one degree, is called its specific heat. From the definition of the unit of heat, it is obvious that in either the English or the French system, the specific heat of water is unity. All others are therefore relative numbers (mostly fractions) referred to water as unity, and are the same in both systems, just as specific gravities are. The *thermal capacity* of a body is its specific heat multiplied by its mass.

Specific heats vary slightly with the temperature, so that the statement that the heat absorbed by a body is proportional to the rise in temperature, is true only for a limited range. The specific heat of a substance is also very different in its different states. For example, the specific heat of ice is 0.45; of water, 1; of steam, 0.5.

The addition of heat to a body, however, does not always raise

---

* The therm is often understood to stand for 100,000 B.T.U., but note that the author has used it as equivalent to 1 B.T.U.

its temperature, nor the abstraction of heat always lower its temperature. Thus, Joseph Black discovered that water when boiling absorbs an enormous quantity of heat without any rise in temperature. Under atmospheric pressure water boils at 100 degrees centigrade, and to convert a gram of it into steam of the *same* temperature, requires the addition of 535 calories, or more than five times the amount of heat required to raise that gram from the freezing to the boiling point. In English measure, water boils at 212 degrees Fahrenheit, and to convert a pound of it into steam of the same temperature requires 970 therms. This heat that must be added to convert a liquid into its vapor, is called the *latent heat of vaporization*. Conversely, to condense steam at 100 degrees centigrade or 212 degrees Fahrenheit to water of the *same* temperature, requires the abstraction of this same latent heat.

The boiling points of liquids are raised by increased pressure, and lowered by diminished pressure. Thus water under two atmospheres (15 pounds gauge or 30 pounds absolute) boils at 250 degrees Fahrenheit; at half an atmosphere (7½ pounds absolute) it boils at 180 degrees; and so on. Boiling points are now conveniently given in the engineer's Steam Tables.

For every pressure (within limits) there is one particular temperature (the boiling point corresponding to that pressure) at which water and steam can, so to speak, live together. The pressure being fixed, if heat is added to such a *mixture,* the water boils. If heat is abstracted, the steam condenses. But the temperature does not change. Both are *isothermal* operations. The steam can only be heated above the boiling point corresponding to the pressure, after all the water has boiled off. It is then said to be *superheated.* Conversely, the water can only be cooled below the boiling point, after all the steam has condensed. In short, to alter the temperature of either, they must be separated. The temperature of the mixture can only be altered by changing the pressure, thus giving a new equilibrium temperature or boiling point. These remarks apply of course to any liquid and its vapor.

The heat that raises the temperature of a body is called *sensible heat* in distinction to latent heat. The amount of heat required to raise a pound of water from the freezing to the boiling point corresponding to a given pressure, is called the *heat of the liquid*. To vaporize the water at this temperature, the latent heat must be added. The sum of these two quantities is called the *total heat* of the steam. Because the latent heat is so large, it is evident that steam contains considerably more heat than water does at the same temperature. For this reason, steam is an excellent medium for the transport of heat, for the purpose, for example, of carrying it into the cylinder of an engine. All that a gas can carry is its sensible heat, and this is small for two reasons—the low specific heat of a gas (only a fraction that of water), and the low density of a gas, by which a huge volume is required to give an appreciable heat carrying weight.

Watt made a number of measurements of the total heat of steam at various pressures, and supposed it constant. Since the heat of the liquid rises with the pressure (because the boiling point rises), this conclusion would require the latent heat (the difference between the two) to diminish. Others—including Carnot—supposed the latent heat to be constant, which would require the total heat to increase continually. The true relations between the three quantities are shown in Figure 8. We see that Watt was much nearer the truth, though none of the surmises were quite correct. The total heat rises slightly at first, then sinks and joins the rising heat of the liquid. The latent heat diminishes, slowly at first, then rapidly, and reaches zero at 705.3 degrees Fahrenheit. This is the *critical temperature* of water, above which it cannot exist as a liquid at all, above which it is a permanent uncondensible gas. Hence none of the curves extend beyond this point. It is the last boiling point and corresponds to a pressure of 3206 psia.

To return to Carnot, since he had to abide by the caloric theory, he was confronted with the problem of explaining how a material substance like caloric could become a source of power. But we have an everyday example in the power of falling water. Water will always flow from a higher to a lower level, but will not run uphill. Similarly, heat will always flow from a higher to a lower temperature, but will not flow up-temperature. Merely falling water, however, is not enough. We must interpose a wheel, the chief function of which is to prevent the free fall of the water, and by opposing a resistance, make it do work as it descends. Similarly a heat-engine prevents the free fall of heat, and makes it do work as it descends in temperature. Said Carnot:

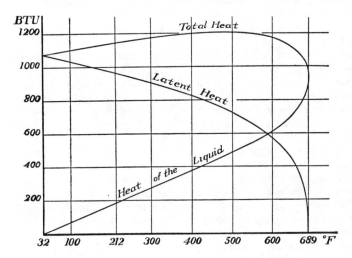

FIG. 8. LATENT AND TOTAL HEATS OF STEAM

The production of motion in steam-engines is always accompanied by a circumstance on which we should fix our attention. This circumstance is the . . . passage (of heat) from a body in which the temperature is more or less elevated, to another in which it is lower. What happens in fact in a steam-engine actually in motion? The caloric developed in the furnace by the effect of the combustion traverses the walls of the boiler, produces steam, and in some way incorporates itself with it. The latter carrying it away, takes it first into the cylinder, where it performs some function, and from thence into the condenser, where it is liquefied by contact with the cold water which it encounters there. Then, as a final result, the cold water of the condenser takes possession of the caloric developed by the combustion. It is heated by the intervention of the steam as if it had been placed directly over the furnace. The steam is here only a means of transporting the caloric.

The production of motive power is then due in steam-engines not to an actual consumption of caloric, but *to its transportation from a warm body to a cold body.* . . . We shall see shortly that this principle is applicable to any machine set in motion by heat.

According to this principle, the production of heat alone is not sufficient to give birth to the impelling power: it is necessary that there should also be cold; without it, the heat would be useless. And in fact, if we should find about us only bodies as hot as our furnaces, how can we condense steam? What should we do with it if once produced? We should not presume that we might discharge it into the atmosphere, as is done in some engines; the atmosphere would not receive it. It does receive it under the actual condition of things, only because it fulfils the office of a vast condenser, because it is at a lower temperature; otherwise it would soon become fully charged, or rather would be already saturated.

Wherever there exists a difference of temperature . . . it is possible to have also the production of impelling power. Steam is a means of realizing this power, but it is not the only one. All substances in nature can be employed for this purpose, all are susceptible of changes of volume, of successive contractions and dilatations, through the alternation of heat and cold. All are capable of overcoming in their changes of volume certain resistances, and of thus developing the impelling power. A solid body—a metallic bar for example—alternately heated and cooled increases and diminishes in length, and can move bodies fastened to its ends. A liquid alternately heated and cooled increases and diminishes in volume, and can overcome obstacles of greater or less size, opposed to its dilatation. An aeriform fluid is susceptible of considerable change of volume by variations of temperature. If it is enclosed in an expansible space, such as a cylinder provided with a piston, it will produce movements of great extent. Vapors of all substances capable of passing into a gaseous condition, as of alcohol, of mercury, of sulphur, etc., may fulfill the same office as vapor of water. . . . Most of these substances have been proposed, many even have been tried, although up to this time perhaps without remarkable success. . . .

Heat can evidently be a cause of motion only by virtue of the changes of volume or of form which it produces in bodies. . . .

Wherever there exists a difference of temperature, motive power can be produced. Reciprocally, wherever we can consume this power, it is possible to produce a difference of temperature. . . . Are not percussion and the friction of bodies actually means of raising their temperature . . . ? It is a fact proved by experience, that the temperature of gaseous fluids is raised by compression and lowered by rarefaction. This is a sure method of changing the temperature of bodies. . . . The vapor of water employed in an inverse manner to that in which it is used in steam-engines can also be regarded as a means of destroying the equilibrium of the caloric. (*I.e.*, of raising heat to a higher temperature.)

Carnot then asked whether the motive power obtainable from a given quantity of heat, the difference of temperatures also being given, "varies with the working substance employed to realize it, whether the vapor of water offers in this respect more or less advantage than the vapor of alcohol, of mercury, a permanent gas, or any other substance." To answer this question, he invites us to consider the operations of a steam-engine, and how they may be perfected.

The first operation is the admission of steam to the cylinder. During this operation, the cylinder and the boiler being in direct communication become essentially one vessel. As the piston advances, the volume of the combined vessels is enlarged. The expansion would cause a drop in pressure, were it not that more water is immediately vaporized and replaces the steam that has entered the cylinder. Thus the pressure is kept up. The vaporization requires the latent heat to be supplied, and this is drawn from the fire. An exactly equal amount of heat (supposing none lost on the way) is carried by the steam into the cylinder. The temperature during the operation remains constant at the boiling point of water for the pressure used. In this way, the engine takes on a load of heat.

The effect would be precisely the same if we put the proper amount of water directly in the cylinder—the piston pressed against it with the boiler pressure—and then put the whole on the fire, keeping it there until the water was completely vaporized. Exactly the same amount of heat would be drawn from the fire as before, and the piston would rise the same amount, being pushed up by the expansion of the water on becoming steam.

Let us then imagine the arrangements shown in Figure 9, which are those of Carnot with some slight later improvements. At A and C are two large bodies kept at the constant temperatures T_1 and T_2 respectively. T_1 is the temperature of the boiler, T_2 is that of the condenser. These two bodies we must imagine so large that whatever amount of heat we abstract from or add to them will not alter their temperatures. We shall call A the heater, and C the cooler. In order that conditions shall be as perfect as possible, let us further imagine that the tops of these bodies and the bottom of the cylinder, are perfect conductors of heat, while the sides of the cylinder and the piston are absolute non-conductors. Between A and C is also a non-conducting stand B.

We now perform the following operations:

1. We place the cylinder with its contained water on A, and

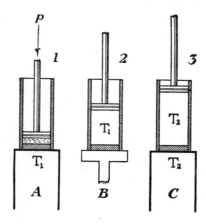

FIG. 9. OPERATIONS OF THE PERFECT STEAM-ENGINE

drawing heat from the latter, vaporize the water at the temperature T_1. The piston rises as already described.

2. We transfer the cylinder to the non-conducting stand B, and, by gradually relieving the pressure on the piston, allow the steam to expand further. Being now surrounded by non-conductors, no heat can escape from the steam, and none can enter it. Under these circumstances its temperature falls, as with any gaseous body. We continue the expansion until the temperature has fallen to T_2. At the same time the pressure will fall from that of the boiler to that of the condenser.

3. We place the cylinder on C, and, pushing the piston down with a constant pressure equal to that of the condenser, liquefy the steam at the lower temperature, and reject its latent heat to C.

The operations which we have just described, said Carnot, might have been performed in the inverse order and direction. There is nothing to prevent our drawing heat from C, forming steam with it at the lower temperature, compressing the steam while out of contact with conducting bodies until its temperature is elevated to that of A, and finally condensing it at the higher temperature and delivering its latent heat to A.

By our first operations we transfer heat from A to C, from a higher to a lower temperature, and thereby gain motive power.

By the inverse operations we *expend* motive power, and thereby lift heat from C to A, just as we pump water uphill. If the operations have been perfect, then by the inverse operations we extract from C the same latent heat that was previously delivered to it, and deliver to A the same latent heat that was previously extracted from it, and in so doing expend the same amount of work that was previously gained by the direct operations.

For, said Carnot, if there existed a more perfect way of deriving motive power from heat than we have described, a method by which more power could be obtained from the same quantity of heat and the same temperatures, then we could divert a *part* of the power delivered by this pluperfect engine, and use that *part* by means of the inverse operations to restore the whole of the heat to the source. By indefinitely repeating the operations, the remaining unused part of the power delivered by the engine could be multiplied without limit. This combination could thus, by simply using the *same* heat over and over again, deliver an unlimited amount of power, without the *permanent* transfer of any heat to a lower temperature. "This," said Carnot, "would be not only perpetual motion, but an unlimited creation of motive power without consumption either of caloric or any other agent whatever. Such a creation is entirely contrary to ideas now accepted, to the laws of mechanics and of sound physics. It is inadmissible. We should then conclude that *the maximum of motive power resulting from the employment of steam is also the maximum of motive power realizable by any means whatever.*"

We can apply the same argument to the water-wheel. By transferring a given quantity of water from a given higher to a given lower level, we can obtain a certain amount of motive power. Now by reversing the buckets, we can scoop up the water from the lower level and deliver it to the higher. If the wheel is perfect, we will in this way expend the same amount of energy as was previously gained, and restore to the source the same amount of water as previously descended. For, if there existed a more perfect way of developing water-power, say a turbine which would develop more power from the same quantity of water and the same descent than our perfect wheel, then we could divert a *portion* of the power delivered by the turbine to operate our wheel with reversed buckets, and restore all the water used to the higher level. We could thus obtain an unlimited amount of power without the *permanent* transfer of any water to a lower level. That would be a perpetual motion, and being inadmissible, we must conclude that the *maximum* motive power obtainable from a water-wheel is also the *maximum* obtainable by any means whatsoever.

Incidentally, we may remark that this is precisely what the perpetual motionist seeks to do. He thinks that some way of lowering a weight differently from that by which it is raised can be found, by which the weight will do more work in descending than is required to raise it again to the same height. Thus he expects by merely moving a weight up and down, to get an unlimited supply of power. Or he thinks to do the same with heat, electricity (as when he sets a dynamo to run a motor, and the motor to run the dynamo), or with some other agency. As a matter of fact, there is a loss both in the descending and in the ascending operations, hence a double loss. Instead of an excess of power being gained, there is a defect that must be supplied, if the combination is to be kept running.

Carnot's conclusion therefore is that the maximum motive power obtainable from a given quantity of heat, depends only upon the temperatures, and not on the working substance used. He says:

We can compare with sufficient accuracy the motive power of heat to that of a waterfall. Each has a maximum that we cannot exceed, whatever may be, on the one hand, the machine which is acted upon by the water, and whatever, on the other hand, the substance acted upon by the heat. The motive power of a waterfall depends on its height and on the quantity of the liquid; the motive power of heat depends also on the quantity of caloric used, and on what may be termed, on what in fact we will call, the *height of its fall,* that is to say, the difference of temperature of the bodies between which the exchange of caloric is made.

We have a right to ask, says Carnot, by what sign it can be known that this maximum is attained. Putting his discussion into somewhat more modern form, it can be said that there are two signs or tests. The first is that all flow of heat should take place between bodies that are at the same temperature. The second is that all temperature changes should be effected solely by the expansions and contraction of the working substance, without flow of heat into or out of it. Obviously the second condition is a corollary of the first, for if all flow of heat takes place without temperature change, then all temperature change must take place without flow of heat.

These two conditions of perfection arise from the fact that wherever a temperature difference exists, motive power can be developed. Hence every such difference that is not so utilized, means that a portion of the available power has not been developed. Suppose, for example, that we put the heater and the cooler in direct contact. Heat will then simply flow from one to the other, but we obtain no motive power. Similarly any lesser drop in temperature by which heat is conveyed from one part of the mechanism to another, means the loss of a portion of the available power. Heat added to the working substance while its temperature is changing, also means a loss, for it is added at less than the topmost temperature, and we lose the power that could have been developed by its fall from that temperature. Similarly, any heat abstracted while the temperature is changing means a loss, for it is abstracted above the lowermost temperature, and we lose the power that could have been developed by its fall to that temperature. Hence, in the perfect engine, all heat must be added at constant temperature, that is, *isothermally at the topmost temperature,* and all must be rejected *isothermally at the lowermost temperature.* In short, the heat flow and the temperature changing operations must be completely separated.

The same rules apply to the perfect water-wheel. All flow of water must take place between vessels that are at the same level, for otherwise we lose the power that could have been developed from the difference of levels. Hence, all water must be taken by the wheel at the topmost level, and discharged at the lowermost. No water must drop into or out of the buckets while they are ascending or descending. All changes of level must be effected solely by the buckets.

It is easy to see that these conditions cannot be fully complied with in the case of an actual water-wheel. But we should endeavor to approximate to them as closely as possible. To determine a flow of water, there must always be a difference in levels. To get the water to and away from the wheel, there must be a slope both in the penstock and in the tailrace. But we can make these slopes gentle by making the channels and the wheel wide, and the speed of turning slow. Again, we cannot fill each moving bucket while exactly at the topmost point of the wheel. We must perforce do some of the filling while the bucket is ascending or descending. Similarly, we cannot dump all the water out at the lowermost point. Finally, it is usually impossible to prevent the water from splashing, spilling, or leaking from the buckets while descending, or some of the water from being swept up by the ascending buckets. But we can aim to make all these losses small.

Exactly analogous defects inhere in every heat-engine. To determine a flow of heat, there must be a fall in temperature. Hence there is always a loss both in getting the heat into, and in getting it out of the engine. But we can make the loss in both cases small by using good conductors for the bodies between which heat is to flow, by making the surfaces in contact large, the contacts good, the distance the heat must travel short, and finally by allowing ample time for the transfer.

Of course the actual steam-engine does not take in heat by conduction through the bottom of the cylinder, as described in Carnot's operations. That would be much too slow. It is carried in by the steam, which has received it from the fire under the boiler where the surfaces of contact are very large. But it would be a mistake to suppose that there is no loss in this process, even if no heat leaks out of the steam-pipe on the way. To determine a flow of steam, there must be a difference in pressure between the two ends of the pipe. Now a drop in pressure of the steam means an increase in its volume—an expansion—and consequent cooling. In fact, the steam-pipe is itself a heat-engine. A small drop in temperature produces the motive force necessary to drive the steam against the friction of the pipe. Similarly, to get the steam out of the cylinder, the back pressure must be higher than the condenser or atmospheric pressure. The steam again expands and cools in the exhaust pipe. We always have to sacrifice a small part of the total temperature drop available to get the steam into and out of the engine.

One might think that these considerations did not apply at all to the modern internal combustion engine, for since the heat is developed directly in the cylinder, there surely can be no loss in getting it in. But this again would be a mistake. To get the explosive charge into the cylinder, it must be sucked in. A partial vacuum must be formed, which means that the piston must do work against the slightly higher atmospheric pressure. In the two-cycle engine, the explosive charge is slightly compressed first, in order that it may shoot into the cylinder the moment the admission port is opened. Indeed, to get the gases into and out of the cylinder consumes from 2 to 5 per cent of the engine's power.

Nor in any engine is all of the heat always taken in at the topmost temperature, or all of it rejected at the lowermost. Indeed, the greatest violators of this rule are the gas-engines, which regularly take in heat on a rising temperature, and reject it on a falling temperature, like a water-wheel that fills its buckets while ascending, and empties them while descending.

Finally, the second rule, that there must be no leakage of heat while the temperature is being changed, is violated by all engines. We cannot build them of non-conductors. On the contrary, we must make them of metal, which is a good conductor. There are two ways, however, in which this leakage loss can be reduced. One is, in the case of the steam-engine, to keep the cylinder hot by means of a steam jacket. If the cylinder could always be kept at the same temperature as the working substance, that would in effect render the metal a non-conductor. But the temperature of the working substance varies. The jacketed cylinder will assume a temperature between the maximum and the minimum, with heat leaking out part of the time and in the rest of the time. However, the leakage is diminished by this means. The other method is to make the expansions and compressions by which the temperature is changed so rapid that there is not time for much heat to leak in or out.

The first rule for approximating perfection thus requires slow speed; the second, high speed. Since the engine must rotate with uniform speed, a compromise has to be effected between these two conflicting demands. Other considerations make high speed desirable, so that we have to approximate to the first condition as best we can by other means than slow speed. Also there is leakage in getting the heat to and away from the engine, so that even these operations cannot be too slow.

One further question remained. Carnot wrote:

> In the waterfall the motive power is exactly proportional to the difference of level between the higher and lower reservoirs. In the fall of caloric the motive power undoubtedly increases with the difference of temperature between the warm and the cold bodies; but we do not know whether it is proportional to this difference. We do not know, for example, whether the fall of caloric from 100 to 50 degrees furnishes more or less motive power than the fall of this same caloric from 50 to zero. It is a question which we propose to examine hereafter.

Carnot examined the question later in his essay. From empirical data, he found that the fall of "caloric" from 100 to 50 degrees yielded in fact less work than the fall of the same caloric from 50 degrees to zero. But he was unable to determine the law according to which this diminution took place, or to explain it. He thought it was due to a large variation in the specific heat of a gas or vapor with its volume, which, we shall see, was one of the false tenets of the caloric theory. However, we have an analogous situation with the waterfall, which Carnot seems to have overlooked. A waterfall ten feet high near the top of a mountain does less work than a similar waterfall near sea level. This is due to the diminution of the force of gravity with the height. We know the law of this diminution. Carnot did not know the law according to which, so to speak, the force of heat diminishes with the temperature height. He left it to his successors to find out, who labored long and vainly at it. They called it *Carnot's function*—a function being an unknown mathematical relation. The problem was not solved until in 1850 Clausius applied the principle of the conservation of energy. It thus turned out that Carnot's difficulty at this point was due to the caloric theory, which here blocked further progress. It was no small mark of his genius that he knew when to stop. He developed his principles only so far as they did not lean too heavily on the caloric theory. He stopped at the point where further advance required a definite assumption as to the nature of heat. And it is significant that precisely at this point he expressed his doubts as to the "solidity" of the caloric theory.

CHAPTER V

THE PERFECT CYCLE

IN DESCRIBING the operations of the perfect steam-engine in the preceding chapter, we left the cylinder standing on the cooler with the steam all condensed at the temperature of this body, and the piston resting on the water with the condenser pressure. Says Carnot:

> If we wish to begin again an operation similar to the first, if we wish to develop a new quantity of motive power with the same instrument, with the same steam, it is necessary first to reëstablish the original condition—to restore the water to the original temperature. This can undoubtedly be done by at once putting it again in contact with the body *A;* but there is then contact between bodies of different temperatures, and loss of motive power. It would be impossible to execute the inverse operation, that is, to return to the body *A* the caloric employed to raise the temperature of the liquid.

In this statement Carnot brings out two important points: First, that the operations of every heat-engine are cyclical. The working substance is always brought back to the original condition, and therefore suffers no permanent change. Second, that a perfect operation is reversible. A flow of heat down-temperature without the development of motive power is irreversible, for the inverse operation would mean to induce the heat to flow back up-temperature without the expenditure of work, which is impossible.

The method above described of completing the steam cycle, corresponds closely to what is done in practice. A feed-pump increases the pressure on the water, and thereby forces it into the boiler, where it is reheated. But, as Carnot points out, the cycle so completed is not reversible and is not perfect. Indeed,

Carnot did not believe that any cycle involving liquids or solids could be carried out in a completely perfect manner. He said that the temperatures of these bodies cannot be changed by compressions and expansions. He pointed out that certain parts of an engine undergo repeated alternations of compression and tension, yet are not heated or cooled thereby. He supposed therefore that a really perfect cycle could only be carried out with a permanent gas. But in this he was mistaken. A perfect cycle can be carried out with *any* working substance, solid, liquid, or gaseous, or any mixture of them, and through any change of state.

To complete the steam cycle in a perfect manner, we must stop the condensation in the third operation before it is quite completed, transfer the cylinder to the non-conducting stand, and there compress the mixture to the boiler pressure. This will also raise the temperature to that of the boiler. If the point at which the third operation was stopped has been properly chosen, this final compression will just condense the remaining steam, and the working substance will be restored to exactly its original condition without the influx or outflow of heat, without the contact of bodies at different temperatures, and therefore in a perfect and completely reversible manner. While this is theoretically possible, practically it is too complicated to imitate, and the saving would not be great. Furthermore, the feed-pump loss can be considerably reduced by heating the feed water before it is returned to the boiler, using for this purpose waste heat which would otherwise have escaped up the chimney.

The complete cycle of a perfect heat-engine thus consists of four operations. During the first and third, heat is taken in and expelled respectively, both operations being isothermal. During the second and fourth, the temperature of the working substance is changed by expansion and compression respectively, without flow of heat. These last two operations we now call *adiabatic,* a word coined by Clausius from three Greek stems meaning literally—no passing through. The indicator card of a perfect, or *Carnot cycle,* thus consists of two *isothermals* and two *adiabatics.*

The same four operations occur in a water-wheel. During the

first and third, the buckets are filled and emptied respectively; during the second and fourth, the elevation is changed.

In actual engines the four operations may be somewhat blurred. The temperature may change while heat is being absorbed or rejected, and heat may flow in or out while the temperature is being changed. Nevertheless, four operations can always be distinguished, which in the main accomplish the four purposes described.

The indicator card of a perfect steam cycle would appear as in Figure 10. Beginning at a, the volume represented by the horizontal distance of this point from the left-hand axis, is that of the initial water in the cylinder, which is at boiler temperature and pressure. By drawing heat from the source, this water is vaporized, and the resulting expansion is represented by the admission line ab, which is both a constant temperature and a constant pressure line. The volume at b is actually about a thousand times greater than that at a, so that b should be a thou-

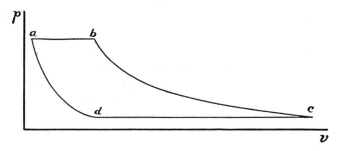

FIG. 10. INDICATOR CARD OF THE PERFECT STEAM CYCLE
(CARNOT)

sand times further from the axis than a. But for convenience it is not made so much in the drawing.

The second operation is the adiabatic expansion bc, by which the temperature and pressure are reduced to those of the condenser. During this expansion, on account of the cooling, some of the steam condenses, a fact that was not known until long after Carnot's time. Hence, even with no loss of heat, as here supposed, there is already a mixture of steam and water at the point c.

The third operation is condensation along *cd*. This again is both a constant temperature and a constant pressure line. At *d*, the condensation is stopped, the point being so chosen that the final adiabatic compression *da*, just condenses the remaining steam, and restores everything to the initial conditions at *a*. The cycle thus consists of two isothermals and two adiabatics, as does every perfect cycle. It is the "Carnot cycle" for steam, for although Carnot did not specifically describe it, he laid the foundations for it.

An indicator card of the steam cycle described by Carnot is shown in Figure 11. It differs from the perfect cycle only in that the condensation along *cd* is continued until at *d* the steam is

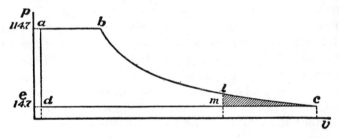

FIG. 11. THE RANKINE STEAM CYCLE

completely liquefied. The pressure is then raised by a feed-pump, and the water reheated in the boiler. This last operation takes place at nearly constant volume, and so is represented by the vertical line *da*. Because heat is added during this operation while the temperature is rising, the ·cycle is not perfect. But, although its efficiency is less than that of the perfect cycle, its capacity is greater, as may be seen by comparing its area with that of Figure 10. Although this cycle was first described by Carnot, it was first applied to actual engines by Rankine, and bears his name. Carnot drew no indicator cards; they were supplied later by Clapeyron and others.

The actual steam-engine follows the Rankine cycle very closely, as may be seen by comparing Figure 11 with the indicator card of Figure 5, page 19. There are two main differences. First, the compression *ca* of the indicator card does not appear in the Rankine cycle. This compression must not be confused with the adiabatic compression of the perfect cycle of Figure 10. The cushion steam in an engine performs the same function in arresting the motion of the piston as would a spiral spring. It does not enter or leave the cylinder, hence does not bring in or carry away heat. It does not belong to the heat cycle at all. It is a part of the mechanism of the engine. In fact, the actual indicator card must be corrected by subtracting the volume of the cushion steam at every point of the diagram, and when this is done, it still more closely resembles the Rankine cycle.

The second main difference is, that in the actual engine the toe of the Rankine cycle is cut off, by some such line as *lm*, Figure 11. (On the indicator card, the corners are also rounded.) The expansion is not carried completely down to the back pressure at *c*, but the exhaust valve opens at *l* at a somewhat higher pressure, which then drops abruptly about *lm*. This is called *incomplete expansion*. It enables the volume of the cylinder to be considerably reduced—to *me* in place of *ce* in the drawing—thus saving weight, space, and expense of the engine. The work obtained is also reduced, but only by the small shaded area of the toe. This amounts in the case illustrated to 10 per cent, while the volume of the cylinder is reduced 30 per cent.

Carnot believed that his cycle could only be properly carried out with a perfect gas. The indicator card of such a cycle is shown in Figure 12. It is composed, as always, of two isothermals and two adiabatics. But while the adiabatics 2 and 4 appear much the same as those for steam, the isothermals 1 and 3 are quite different. They are no longer constant pressure or horizontal lines, but down-sloping pressure-dropping lines. They are simply not so steep as the adiabatics. It is easy to see that this must be so. When a gas expands without inflow of heat, its temperature falls. To carry out the expansion isothermally, heat must be added to prevent this temperature fall. The added heat expands or increases the pressure of the gas, and so prevents the pressure from dropping as rapidly as it otherwise would. The isothermal is therefore less steep than the adiabatic.

When a gas expands isothermally, the pressure varies inversely as the volume. That is, when the volume is doubled the pressure is halved, and when the volume is halved the pressure is doubled, and so on. This simple law was discovered by Robert Boyle in

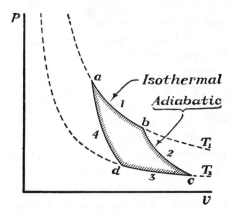

FIG. 12. THE PERFECT GAS CYCLE

1662. It enables us at once to draw the isothermals of a gas.

In 1802, Gay-Lussac discovered that if a gas is heated while its pressure is kept constant, it expands in proportion to the rise in temperature. In short, the expansion is uniform. On the other hand, if a gas is heated at constant volume, the pressure increases in proportion to the temperature rise—also uniformly. Furthermore, he found that all the so-called permanent gases expanded or increased their pressures, as the temperature rose, at the same rate. They all had the same temperature coefficients. Physically then, all the permanent gases behave much alike. They differ only in their densities and specific heats.

Although it has been shown by the more accurate measurements of Regnault and others that these laws of Boyle and of Gay-Lussac are not exactly obeyed by gases; nevertheless, for temperatures at which the gases are far removed from their liquefaction points, the approximation is very close. Since in the case of engines we are always dealing with high temperatures, we commit but little error in assuming these laws to be obeyed exactly. And this is very convenient, for then one formula covers all gases, and that formula is very simple. We therefore assume an ideal or *perfect gas,* which is simply one that obeys the laws of Boyle and of Gay-Lussac exactly.

The operations of the gas cycle are exactly the same as those of the steam cycle, only we start at *a*, Figure 12, not with water in the cylinder, but with a quantity of gas at the maximum temperature and pressure of the cycle. Along *ab*, we absorb heat isothermally at this highest temperature; along *bc* we expand the gas adiabatically until it sinks to the lowest temperature; along *cd* we compress the gas isothermally and reject heat to the cooler; along *da* we compress the gas adiabatically until its temperature and pressure are raised to the initial values.

The complication of a change of state is avoided in this cycle. The substance is always gaseous. And since the laws of a perfect gas were well known even in Carnot's time, it was much more amenable to calculation than the steam cycle. However, the specific heats of gases were not well known; the quantity of heat absorbed during an isothermal expansion had never been directly measured; and the fall in temperature on sudden expansion had only been occasionally and very roughly measured. Carnot was the first to calculate these last two quantities from such data as were then available. The exact calculation, however, of the quantities involved even in this simple cycle, had to await the work of Clausius in 1850.

The outstanding feature of the Carnot cycle that distinguishes it from all others is its *reversibility,* in the sense that the same work is required to restore the heat to the source as was obtained from its descent. No other cycle is reversible in this sense, although every cycle can be *reversed.* But in other cycles *more* work is required to reëlevate the heat than was obtained from its descent. Each operation of the Carnot cycle is also reversible. This is true only of isothermals and adiabatics. While any other path can be retraced, the amounts of work involved in the direct and in the inverse operations are not the same. Since no actual processes are ever truly isothermal or adiabatic, it follows that all natural processes and all actual cycles are *irreversible.* There are irretrievable losses in both the direct and the inverse processes. We bring the working substance of an engine back to its initial state, only by making up the losses, by supplying additional heat or work from elsewhere. When *all* the bodies concerned are considered, a previous state of the universe can never be exactly duplicated. There is always a residual one-wayness about events. However perfectly some of them may seem to repeat themselves, there are also continuous changes going on that cannot be undone. The parts of a machine wear out. A compressed spring, when released, does not restore exactly the work spent in compressing it; some is lost in external and internal frictions. History may repeat herself, but she never does so exactly. Only ideal processes are perfectly reversible.

CHAPTER VI

CARNOT'S RECOMMENDATIONS FOR THE IMPROVEMENT OF ENGINES

SINCE the amount of motive power obtainable from a given quantity of heat depends upon the temperature fall, the most effective way to increase the efficiency of engines, Carnot pointed out, is to widen the temperature limits between which they work.

Now the lowest temperature at our disposal is that of the surroundings. If we happened to have a body at a lower temperature, it is obvious that the continual addition to it of heat would soon raise its temperature to that of the surroundings, and our advantage would only be temporary. The surroundings alone provide a reservoir so vast that no amount of heat that we can pour into it will raise its temperature. In practice, we cannot even get quite down to this level, because a final temperature drop is required to carry the heat away from the engine. In the case of a condensing steam-engine, the bottom temperature for the engine is the top temperature of the condenser. The latter can be reduced by increasing the supply of cooling water, but larger circulation pumps will then be required and more power must be drawn from the engine to run them. Hence we cannot go too far in this direction. A considerable volume of water is required in any case, and cooling water is sometimes scarce and expensive—as in cities. For this reason, the same water is sometimes used over and over. After being warmed in the condenser, it is led to the top of a cooling tower, where it dribbles down over a number of horizontal shelves, cooling as it descends, and is then returned to the condenser.

The top temperature of a condenser is seldom less than 100 degrees Fahrenheit, and usually much higher. Indeed the cistern in which the water resulting from the condensation is collected, and from which it is pumped to the boiler, is called, in engine-room parlance, "the hot well." Yet for the engine it is the "cold body."

If the engine exhausts to the atmosphere, its bottom temperature is 212 degrees. The great advantage of the condenser is, that it enables this bottom temperature to be lowered.

Since we cannot greatly lower the bottom temperature of a steam-engine, the only alternative is to raise the top temperature. This means higher pressure, for the top temperature is determined by the boiling point of water under the pressure used. But higher pressure, Carnot pointed out, is of little advantage unless the expansion is correspondingly increased. The steam should expand from the boiler pressure down to that of the condenser. Its temperature will then also sink to that of the condenser.

Moderately high pressure engines—around fifty pounds—existed in Carnot's time, and higher pressures had been experimented with. But the expansion used was little or no more than that of the low pressure engines. Consequently they failed to realize any great gain in efficiency. The typical cycle of such an engine is *abefd,* Figure 13. The exhaust valve opening at *e,* the rest of the possible expansion *ec* took place in the exhaust pipe, and the work which this expansion produced, represented by the considerable area *ecf,* was lost.

How little this matter was understood, is shown by an engine of the time mentioned by Carnot. It was built by a Mr. Perkins, a celebrated London mechanician in 1823, and used the then unheard of pressure of five hundred pounds per square inch. But steam was admitted for the *full length of the stroke.* There was no expansion at all. Its cycle was *abgd,* Figure 13. The whole of the work represented by the area *bcg* was thrown away. The exhaust valve opened at *b,* with the steam at the full boiler pressure and temperature. It would seem that the folly of letting steam at five hundred pounds pressure shoot out into the air without doing anything useful, should have been apparent even to a mere mechanician, however unaware he might have been

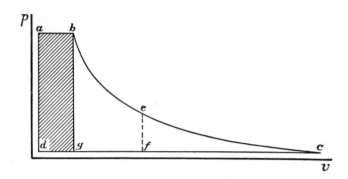

FIG. 13. HIGH PRESSURE STEAM CYCLES

of the thermal atrocity of letting heat drop abruptly from 470 to 212 degrees without making any use of this fall. Carnot's comment on this engine was: "The machine of Mr. Perkins seems not to realize the hopes which it at first awakened. It has been asserted that the economy of coal in this engine was $\frac{9}{10}$ above the best engines of Watt (using 7 pounds!), and that it possessed still other advantages. . . . These assertions have not been verified."

To carry out a large expansion in a single cylinder would require a very early cut-off, and this for many reasons both thermal and mechanical, is impractical. But by dividing the expansion among two or more cylinders, this evil of an early cut-off, we have seen, can be avoided. Hence high pressure with large expansion means compounding. Compound engines had long been invented, but they too used little more expansion than that employed by the single cylinder engines. One of the main objects of high pressure had been to get rid of the condenser. But Carnot pointed out that to get the full benefits of high pressure, the condenser should be retained.

Furthermore, because of the working of the "Carnot function," more is to be gained by lowering the bottom temperature, than by raising the top temperature a like amount. Also, to lower the bottom temperature, say 50 degrees from 212, we only have to lower the bottom pressure 10 pounds, while to raise the top temperature 50 degrees, say from 281 degrees, which is the boiling point at 50 pounds absolute pressure, we have to raise the pressure 54 pounds. By the latter operation, we gain 68 per cent in power, by the former 80 per cent, for the same heat employed in each case. These relations are shown in Figure 14. The central clear portion is the cycle of an engine taking steam at 50 pounds absolute pressure, and exhausting to the atmosphere. Incidentally, this cycle yields the same amount of work as a Watt engine using 7 pounds above the atmosphere, and exhausting to a condenser of 5 pounds absolute. The shaded areas above and below show the portions added to the cycle by raising the top temperature 50 degrees and by lowering the bottom temperature the same amount, while the figures in the areas show the relative amounts of work that each gives.

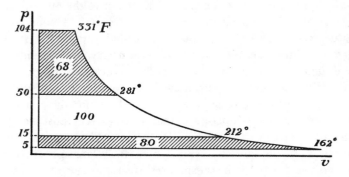

FIG. 14. EFFECT OF RAISING OR LOWERING THE STEAM PRESSURE

This diagram may also represent a triple expansion engine. The upper shaded area would be the indicator card of the high pressure cylinder. Its exhaust line, at 50 pounds, becomes the admission line of the intermediate cylinder, whose card is the middle area. The exhaust line of the latter becomes the admission line of the low pressure cylinder, whose card is the bottom area. Again we may note that this last cylinder must be as large as though the whole expansion had taken place in it alone. If that were so it would have a very early cut-off, in this case at about one-eighth stroke, while with three cylinders, the cut-off in each occurs at about one-half stroke. In practice, the toe of each diagram would be slightly truncated and the corners rounded, and there would be a small difference in pressure between the exhaust of one cylinder and the admission of the next, in order to transfer the steam.

Some engineers of Carnot's time thought that a remarkable advantage would be gained by using, instead of water, some liquid having a lower boiling point. This idea crops up every now and then even today. One claim is that not so much heat is wasted in raising the liquid to the boiling point when starting cold. This of course is true, and may be of some importance in a plant that is frequently shut down. But the main idea has always been that, since the vapor of alcohol exerts a higher pressure than the vapor of water does at the same temperature, more power would be developed. Thus, at 212 degrees steam exerts a pressure of 15 pounds and just balances the atmosphere, whereas at this temperature alcohol vapor exerts 33 pounds. At 300 degrees steam exerts 67 pounds, alcohol vapor 141. At 469 degrees steam exerts 508 pounds, alcohol nearly 1000 pounds. In these higher pressures the engineers saw more push, hence more power. But instead of an advantage they would only constitute a fresh obstacle to overcome, said Carnot. It is not the top pressure, but the top temperature that counts. We must become thermally, instead of mechanically minded. The required push can always be obtained by using a sufficiently large piston, since the push is equal to the steam pressure times the piston area. The principal defect of steam, he said, is its excessive pressure at an elevated temperature, which defect exists still more strongly in the vapor of alcohol. If, for example, we may use a pressure as high as 1000 pounds, then by the use of steam we can reach a temperature of 545 degrees. By the use of alcohol vapor we could reach only 469 degrees, and this cannot be exceeded, no matter what pressure is applied, because it is already the critical temperature of alcohol. Carnot said that, quite on the contrary, "If we could find an abundant liquid body which would vaporize at a higher temperature than water, . . . which would not attack the metals employed in the construction of machines, it would undoubtedly merit the preference. But nature provides no such body."

Nature does provide such a body, but it was not very abundant in Carnot's time. That body is mercury. At atmospheric pressure, it boils at 674 degrees Fahrenheit. At 960 degrees, the gauge shows only 125 pounds. This is 70 degrees above the critical temperature of water (at which the pressure is nearly 3000 pounds), hence in a region where water cannot be used at all. An engine of 30,000 horse-power was built by the General Electric Company, using mercury vapor at the above temperature and pressure. While it would be theoretically possible to expand the mercury vapor until its temperature had sunk to nearly that of the surroundings, its volume would then be so enormous that a condenser as large as an auditorium would be required. Hence the vapor is only partially expanded in a turbine, and the still very hot exhaust mercury vapor is then used to raise steam at 400 pounds pressure. The latter is used in another turbine and condensed in a vacuum in the usual way. The condenser of the mercury engine is thus the boiler for the steam-engine.

By this combination a total drop of 960 − 100 = 860 degrees is obtained. Some 200 degrees must be sacrificed, however, to transfer the heat from the mercury to the water, so that the net drop is 660 degrees. Even so, this is double that obtained from the steam part of the plant, already a highly efficient engine. Thus, by superimposing the mercury engine on the steam-engine, the economy is doubled. This plant was in fact one of the most efficient in the world when it was built. Thus after more than a hundred years, Carnot's suggestion of using a liquid of higher boiling point than water received a magnificent application.

The greatest loss in a steam plant, however, Carnot pointed out, is due to the enormous drop in temperature that occurs between the fire and the water in the boiler. This drop may amount to 2000 degrees, and represents an immediate and huge loss of available motive power. It is from five to fifteen times the drop that is utilized in the engine. It is as though with a waterfall one hundred feet high, we used only the last few feet for power. Yet this loss is unavoidable in any form of vapor engine, because the top temperature for the engine is the boiling point of the liquid under the pressure used. No liquid can be found whose boiling point is sufficiently elevated under a reasonable pressure to approximate that of the fire. Even the mercury-steam combination, above mentioned, utilized only a third of the total drop.

The only way to avoid this loss, Carnot said, is to burn the fuel inside the cylinder. Then the temperature of the fire itself is available, and none of the heat escapes up a chimney. And, since air is the only abundant gas that supports combustion, it is the proper working substance for such an engine.

Explosion engines had been attempted in Carnot's day, but without much success. However, Carnot believed that the difficulties could be overcome, and saw in this type of engine the promise of a great future increase in economy. How well he foresaw is attested by the fact that the internal combustion engine is today our most widely used prime-mover. Carnot further pointed out that for the best results in such an engine the air should be compressed before the fuel is ignited. Nevertheless, engineers were completely ignorant about this matter when the first commercial gas-engines appeared in 1860. Precompression was not used until 1876 in the Otto engine, fifty years after Carnot had indicated the need of it.

Carnot, in fact, pointed out every direction (except one, which we will discuss later) in which major improvements in the economy of heat-engines could be made, and afterwards actually were made, though often belatedly and almost accidentally. Yet he realized that efficiency is not the only virtue an engine can possess. He said: "We should not expect ever to utilize in practice all the motive power of combustibles. The attempts made to attain this result would be far more hurtful than useful if they caused other important considerations to be neglected. The economy of the combustible is only one of the conditions to be fulfilled in heat-engines. In many cases it is only secondary. It should often give precedence to safety, to strength, to the durability of the engine, to the small space which it must occupy, to small cost of installation, etc." And if engines had been applied to any extent to vehicles in his day, he would undoubtedly have added the consideration of weight. The task of the engineer, he said, is properly to balance these different and often conflicting aims in accordance with the use for which the engine is designed.

These remarks are fully applicable today. A locomotive, for example, is not a particularly efficient engine. It is of the single expansion non-condensing type, and uses only moderate pressure. Double expansion and superheat (to be explained later) have been used to some extent to improve the economy, but are not the general practice. A locomotive must have large power in proportion to weight and space; it must be simple and reliable —a breakdown in a desolate place is both inconvenient and expensive; it must have flexibility, *i. e.,* a large range of power and speed; it must be rugged, and not too costly to replace. Economy of fuel is secondary. In fact, the whole fuel bill of a railroad is about 6 per cent of the total expenses, so that if half the coal were saved, the costs would be reduced by only 3 per cent.

Similarly in factories, or wherever the fuel bill is a small part of the total costs, highly efficient engines are unnecessary and even undesirable, because of their greater complexity and consequent greater likelihood to break down—and a shutdown in a factory is usually expensive. It would be absurd, for example, to equip an excursion boat with highly efficient and costly engines, when the boat is only used three months of the year, and even then is tied up to the wharf most of the time. In the commercial world, it is primarily not the thermal efficiency of the engine that counts, but the cost of power, and this is divided between first cost and maintenance. Where fuel is cheap, there is no great need of economy.

When the sole business of a company, however, is the manufacture and sale of power, then no reasonable expense or pains should be spared to make the plant as efficient as possible. In fact, it is with the power companies that we find our most efficient engines today, and every possible device employed that can in any way promote economy.

Carnot was carried off by the cholera epidemic that swept Paris in 1832, when he was but thirty-six years of age. It is interesting to speculate upon what might have been the course of events, if this brilliant young genius had been spared a few more years. From a notebook he left behind, we know that he was just on the verge of that greatest discovery of the nineteenth century, the conservation of energy. Had he lived to complete his studies, the progress of science and particularly that of engineering would undoubtedly have been advanced by some decades. But even his work on the motive power of heat became little known during his lifetime. It was described in 1834 by

Clapeyron in a physical journal, and provided with indicator diagrams. It was largely through this paper that Carnot's work became known to physicists. Somewhat later William Thomson, then a very young man, enthusiastically adopted Carnot's ideas, and pointed out their importance. From 1850 to 1860 Clausius and Thomson combined the theory with the conservation of energy, and worked out its mathematical details. Rankine, Hirn, and Zeuner worked out its methods of application. Yet it was not until 1870 that steam pressures in general use began to rise above 60 pounds, not until 1880 that triple expansion condensing engines in accordance with Carnot's recommendations began to appear. Fifty years were required to get the theory into practice.

CHAPTER VII

ANTIQUITY OF THE CONSERVATION
IDEA

JEVONS[1] wrote: "Science arises from the discovery of Identity amidst Diversity." Helm[2] said that the conservation and transformation ideas are as old as thought itself. Stallo[3] said that the doctrine was "coeval with the dawn of human intelligence." Indeed, the idea that behind the changing world we perceive there is an unchanging order that we do not at once perceive, that all apparent change is but the rearrangement of unchanging parts in accordance with fixed rules, marks the beginning of a scientific and naturalistic view of things, as opposed to a supernaturalistic view, in which magic, animism, and caprice hold sway. According to the latter view, anything can happen. Anything can be transformed into anything else—without due process of nature—a man into an animal, lead into gold, etc., and if we are to believe the ancient tales, these things were of common occurrence in the past. Science does not admit these magical transformations. It demands a process according to fixed rules. It *limits possibilities*. It defines a realm of the impossible.

The first fruit of the scientific search for the "invariants" of nature was the doctrine of the indestructibility of matter, particularly of those small bits of it called atoms. It was the atomist Democritus who said, in the fifth century B. C.—nothing comes from nothing, and nothing can become nothing. And his fol-

1 W. S. Jevons, *The Principles of Science* (Dover, 1958), p. 1 [1st. ed., New York, 1874.]

2 Georg Helm, *Die Energetik* (Leipzig, 1898), p. 5.

3 J. B. Stallo, *The Concepts and Theories of Modern Physics* (4th ed., New York, D. Appleton and Company, 1900), p. 68. [1st ed., 1881.]

lower Epicurus added—for otherwise everything can come from everything—which is merely a denial of magic.

But the conservation of matter was not enough. The atomists attributed eternity also to the *motions* of their atoms, though not of course to each individually. One might come to rest, but somewhere else another was moving. Never could all be brought to rest at one time. Motion was therefore something that could be indefinitely redistributed, but never completely annihilated. All the phenomena of the universe consisted of redistributions of matter *and* of motion.

This idea that motion as well as matter is in some sense conserved, persisted through the ages. But attempts to give it a precise formulation met with difficulties.

A body cannot set itself in motion. Motion must be communicated to it from without. Nor can it, once moving, destroy its own motion. A resistance must be opposed, by which its motion is absorbed by another or by innumerable other bodies. This is the essence of the law of inertia, the most fundamental law of mechanics, as laid down by Galileo at the beginning of the seventeenth century. It is nothing other than a denial of perpetual motion, in the sense of a continual creation of power. A body, or a system of bodies, cannot continually move against a resistance, cannot continually communicate motion to other bodies, that is, *do work*, unless it continually *receives* motion from elsewhere. In some sense then, motion is indestructible and perpetual. But in what sense? How are we to measure and sum up all the motions in the universe, so as to show *what quantity* is conserved?

In the case of matter, it was easy to point to the atoms as the elements conserved. But what, in the case of motion, is the element conserved? Motion is a complex thing. Velocity does not consist of speed alone (miles per hour), but involves also direction. Motion may be communicated to a body by the impact of another. But the effect depends not only on the velocities but also on the sizes, or more exactly on the *masses* of the colliding bodies. Thus a large body striking a smaller one will communicate to the latter a greater speed than it, the large body, loses. Speed is evidently not the element conserved. Indeed, it is possible to strike a moving body in such a way that only the direction of its motion, but not its speed, is altered. Yet the striking body loses speed in doing this. Speed has been paid for a mere change in direction. It does not seem like a fair exchange. But nature makes no unfair exchanges. There must be some way in which the accounts balance.

In 1668, Wallis and Huygens, studying impact, found that if we multiply the mass of the one body by the speed which it loses *in a particular direction on impact,* this product will be equal to the mass of the other body multiplied by the speed which the latter gains *in that same direction.* The product, mass times velocity *(mv),* is therefore the quantity conserved when motion is transferred from one body to another by impact. Since it seemed to be a measure of motion, it was called the *quantity of motion.* We now call it the *momentum;* and what we have described is the law of the conservation of momentum. Newton showed later that this law holds for any system of bodies, and regardless of whether the motion is transferred by direct impact or by the action of forces at a distance.

Descartes, anticipating the results of Wallis and Huygens, drew the conclusion that the total quantity of motion in the universe, defined as the sum of the products of each mass into its speed, *regardless of direction,* was constant. On this assumption he built his famous vortex theory—now only of historic interest as the first attempt at a completely mechanical model of the physical universe. Descartes mistook the import of the momentum law, which, indeed, had not then received the more general formulation later given it by Newton. The law really signifies that if a system as a whole is at rest, then there is on the whole no more motion in any one direction than there is in the opposite direction. This motion in the opposite direction must be given a *negative* sign. When that is done, the oppositely directed motions cancel out, and we find that the total momentum of the system, in any direction we may select, is *zero.* But the motions among the parts, and the sum of those motions (reckoned in the Cartesian manner), may be anything whatever, and may vary continually. This simply means that such a system (which might be the universe) is itself an inert body, and cannot by any interaction of its parts set itself as a whole in motion, nor, if it should happen to be moving, can it alter that motion.

For example: the solar system is moving as a whole in a certain direction in space. No actions within that system and confined to it can alter its motion. If the solar system should collide with another system, or come near enough to affect it by gravitational force, any motion which it communicated to the other system in any direction would be detracted from its own motion in that direction. If the universe as a whole is at rest, there must be right now enough motion of other systems in the opposite direction to that in which the solar system is moving, to balance the latter's motion.

The sum of the internal motions of a system, the sum of the mv's regardless of directions, is therefore not a constant. It is not the "invariant" of motion that we are seeking.

Huygens showed further that when perfectly elastic spheres collide, another quantity is conserved, namely, the sum of the products of the mass of each body into the *square* of its speed. This quantity, mv^2, Leibnitz called the *vis viva* or living force. We now call $\frac{1}{2}mv^2$ the *kinetic energy*. This is not a directed quantity, so that the total kinetic energy of a system is the sum of the kinetic energies of all its parts, regardless of the directions in which they are moving. Leibnitz pointed out that Descartes should have used this quantity mv^2 instead of his mv. But the total kinetic energy of a system is conserved only when the latter is completely isolated from all other systems, and is composed only of perfectly elastic spheres between which no distance forces are exerted.

Attempts have been made to reduce the universe to such a system—to replace all action at a distance by impact, to explain gravitation by means of a connecting medium, to show that all potential energy is really kinetic and due to hidden motions. But it has never been possible to carry through such a scheme. It has never been possible to get rid of the distance forces completely. The total energy of a system is always partly kinetic (due to the motions of its parts), and partly potential (due to forces between bodies which are separated). Hence the law of the conservation of kinetic energy, which at times has been held to be universally true, holds only for special cases.

When two attracting bodies approach each other in obedience to their attraction (such·as a raised weight that falls toward the earth), the potential energy due to their separation and attraction diminishes, and kinetic energy, represented by increased velocity, takes its place. It was shown by Huygens that, in the absence of all resistances, the kinetic energy appearing is exactly equal to the potential energy disappearing. In a more general way it was shown by Lagrange that in any completely isolated system of bodies, moving in obedience to their mutual attractions and repulsions, the sum of the potential and kinetic energies remains constant—provided no complications occur. This is the law of the conservation of mechanical energy.

The solar system is an almost perfect example. As the earth in its elliptical orbit approaches perihelion—the point nearest the sun—its speed increases. Its potential energy diminishes with the diminished distance, while its kinetic energy increases. As the earth recedes toward aphelion, its speed diminishes. Its kinetic energy diminishes, while its potential energy increases. Always the sum of the two remains the same.

If the universe were a purely mechanical system, this law would be universally true, and our search would be ended. But complications occur. One of the most distressing of them is the following. Suppose a system consists of two equal inelastic spheres rushing directly toward each other with equal speeds. Then after collision, both bodies are completely at rest. The entire mechanical energy of the system, potential and kinetic, as well as the total quantity of motion, disappear completely.

This was the sort of difficulty that had to be surmounted before a universal conservation law concerning motion, as a counterpart to the conservation of matter, could be set up, and given exact quantitative formulation. We here arrive at an *impasse,* and must follow up another line of approach.

CHAPTER VIII

HEAT AS A FORM OF MOTION— BACON, 1620

THAT heat and motion are intimately connected was observed in the earliest times. In the restlessness of fire, the Greeks saw the essence of mobility. Its constant upstreaming indicated to them that the sky was its natural abode. The heavenly bodies were of a fiery nature, and therefore mobile. Some thought that fire or a fiery essence was the origin of all things, or at least of all life and movement. The sun's rays sustained life. Life itself seemed to be a sort of fire. It kept the body warm, and when the flame of life was extinguished, the body grew cold. On the other hand, it was observed that motion can produce heat. Thus Plato[1] said: "For heat and fire which generate and sustain other things, are themselves begotten by impact and friction: but this is motion. Are not these the origin of fire?"

Thus heat produces motion, and motion produces heat. The former we see in the draughts of our chimneys, in the winds, in the crinkly atmosphere above a hot surface, in the ebullition of boiling liquids, in the expansion of bodies—particularly that of air. We have seen Hero utilize the latter and the steam produced by heat to set mechanisms in motion. Heat by fusing a solid renders it liquid and mobile, and by vaporizing the latter makes it still more mobile. That motion produces heat, we see, as Plato mentioned, in friction and impact. The savage produces fire by friction. It can also be produced by the sudden compression of air, as in the German pneumatic tinder boxes mentioned by Carnot.

---

[1] Quoted by John Tyndall, *Heat as a Mode of Motion* (6th ed., New York, D. Appleton and Company, 1883), p. 33. [1st ed., 1863.]

With all these everyday evidences of the close connection between heat and motion, it was but a short step to the conclusion that heat *is* a form of motion. The first to take this step appears to have been Francis Bacon, who set forth his ideas and method in his *Novum Organum* of 1620. Since he has been so widely proclaimed as the father of the inductive sciences, and the discoverer of their true method, and seems indeed in this particular instance to have anticipated in a remarkable manner an important scientific discovery of much later date, we shall examine his ideas and method in some detail.

Bacon undertook the investigation of heat not to add to scientific knowledge, but to show how a scientific investigation should be carried out. The important thing was not the result but the method.

It was Bacon's belief that the object of science was to discover the inner natures, essences, or "forms" of things. His aim was still essentially Aristotelian and scholastic. Only the method was different. Observation and experiment were to replace pure reason. From large masses of particulars, the mind was to ascend by induction to the "universal affirmation." That accomplished, the research was ended.

Bacon believed that the process of research could be so mechanized, that it could be carried out by men of little wit, the services of creative genius thus dispensed with, scientific discovery made easy, and placed upon a basis of mass production. He also believed that he had discovered this remarkable method, and set himself diligently to describe it. As an illustration, he gave a sample research into the "nature which is heat."

The first thing to do, he directed, was to collect all the information available concerning the "nature" to be investigated, and to make experiments where information was lacking. These "instances" were then to be drawn up in three great tables, which he called the *Affirmative,* the *Negative,* and the *Comparative* instances. The first table was to include all those cases in which the "nature" appeared; the second, all those in which it did not appear when one might reasonably expect that it would appear; the third, those cases in which it appeared in varying degrees.

A glance at the lists which Bacon drew up for the "nature"

heat, would appall a really scientific man of *any* age. Included in the first list were such statements as that pepper and mustard are hot, that wool and all shaggy substances are warm. In the second, we have that moonbeams are cold despite their resemblance to sunbeams which are warm. In the third, we are told that fire burns more hotly in cold than in warm weather. It is evident that the tables contain a mixture of fact, fable, hearsay, personal opinions, prejudices, and even of superstitions, with no attempt to sift the true from the false. But Bacon had great faith in *system*. He believed that "Truth emerges more readily from error than confusion."

Having drawn up his lists, the investigator was then to cast his eye over them and pick out the common element. As a help in this process, Bacon drew up a fourth or *Rejection* table, in which were listed all the trial generalizations, and the reasons for rejecting them. This tedious process was to be continued until at last the "complete affirmation" was attained, with which the research was ended.

Proceeding in this way, Bacon, after several rejections, announces as his final conclusion that heat is motion. And he is very emphatic about it. He declares that heat is not the *result* of motion, "but that the very essence of heat, or the substantial selfe of heat, is motion and nothing else." A stroke of genius? Perhaps! But let us not forget that Bacon's method was supposed to dispense with strokes of genius. It claimed to be automatic, to be operable by a corps of clerks, statisticians, or machine tenders. Let us examine his conclusion a little more in detail.

Bacon realizes of course that not *all* motion is heat, but only some particular kind. Summarizing his four statements, he described this particular motion as follows: Heat is an outward, upward, expansive motion, not uniform or of the body as a whole, but of the small parts of it, which are perpetually hurrying, striving, struggling, but at the same time always repressed and reflected. This motion takes place "not in the very minutest particles, but rather in those of some tolerable dimensions."

That is nothing other than the motion of boiling water, of flickering flames, of upward streaming convection currents, of the expansion of a body. It is a composite of the chief *effects* of heat, and therefore cannot possibly *be* heat. No one not befuddled by a system could thus mistake the *effects* of an agent for the agent itself.

Bacon similarly defines cold as a "contracting and condensing motion." Thus cold is also a mode of motion, and is objectively as real as heat. That certainly was no contribution to science.

Bacon admits that boiling liquids and fire were his "conspicuous instances," and indeed his description scarcely fits anything else. How does it apply, for example, to the hotness of mustard, to the warmth of shaggy substances? How does it explain why a fire burns more briskly in winter than in summer—supposing that it does—or why moonbeams are cold, etc.? Bacon in fact is scarcely beyond Plato in his conclusion, while he is much behind him in his circuitous method. It was needless to cast up a hundred instances when he only paid attention to two of them. His rejections were not so much of generalizations that did not fit all the instances as of instances that did not fit his final conclusion. This was his method of making truth emerge from error.

Bacon's method is essentially that of pure induction. Logicians know that unless an induction is perfect or complete, that is, unless *all* the pertinent facts are enumerated, the conclusion cannot be certain. Bacon knew it too. He inveighs against hasty generalization. Yet all he recommends is a larger number of instances, and a particular treatment of them. His method is merely one of *less hasty* generalization.

Moreover, Bacon overlooks the possibility of two or more generalizations perfectly fitting all the facts, as often happens. His method provides no means of deciding in such a case. The perfection of an induction is no guarantee of its factual truth. Whether based on few, on many, or on *all* the relevant facts, it is always tentative. It is an hypothesis. It must be verified by testing its consequences. This is the method of Galileo—the method of experimental science. Bacon took only the first step, and a faltering one at that.

It is now so incontestable that the method of Galileo is the method of experimental science, that apologists for Bacon have combed his writings for passages that would seem to indicate that this was the method he attempted to describe. In all, only four passages, to my knowledge, have been found, which, by any

stretch of plausible misinterpretation and lifting from the con-
text, could be so construed.[2] On the other hand, there are plenty
of passages where Bacon unequivocally opposes the method of
Galileo, ridicules those who practised it, and rejects their results.[3] The whole tenor of the book is in this key. Besides, if
Bacon had believed in the method of Galileo, he would have
applied it in his model research. Had he done so, had he
honestly attempted to verify his "complete affirmation," he would
speedily have had to reject it.

Let us then make the tests for him. If heat is a seething motion of parts of tolerable size, such as we see in boiling water,
then where this motion is greatest, there also the heat should be
greatest. But we see this motion not at all in red-hot iron. Yet
the iron is unquestionably hotter, for it will quench in boiling
water. Again, water a few degrees below the boiling point is just
as quiescent, as far as these gross motions are concerned, as water
a few degrees above the freezing point. Yet it is undeniably
hotter, as can be tested by the finger. Oil has a higher boiling
point than water. Hence at a temperature much above that at
which water exhibits the wildest commotion, oil is perfectly

[2] The four passages are as follows:
Axioms properly and regularly abstracted from particulars easily point
out and define new particulars, and therefore impart activity to the sciences.
Novum Organum, Book I, § 24.
The real order of experience begins by setting up a light, and then
shows the road by it, commencing with a regulated and digested, not a
misplaced and vague course of experiment, and thence deducing axioms,
and from those axioms new experiments. *Ibid.*, § 82. This passage is quoted
by Durant in *The Story of Philosophy*, but somewhat differently translated
from the Latin than the above.
In forming our axioms from induction, we must examine and try whether
the axiom we derive be only fitted and calculated for the particular instances from which it is deduced, or whether it be more extensive and general. If it be the latter, we must observe, whether it confirm its own extent
and generality by giving surety, as it were, in pointing out new particulars.
Ibid., § 106.
The signs for the interpretation of nature comprehend two divisions, the
first regards the eliciting or creating of axioms from experiment, the second
the deducing or deriving of new experiments from axioms. Book II, § 10.
[3] Bacon rejected the vacuum, the conservation of matter, atoms, the
Copernican theory, Gilbert's discovery of the earth's magnetism (called him
an alchemist), contradicted Galileo's mechanics (insisted that heavier bodies
fell faster), inveighed against mathematics, specialization, and the breaking
away of the sciences from philosophy.

quiet. Alcohol boils at a lower temperature. Hence it is violently agitated at a temperature for which water is perfectly quiet. These are all tests that Bacon could have made. The boiling points could have been compared by means of Galileo's air thermometer. Bacon was in fact acquainted with this instrument, for he describes it. He could perfectly well, then, have applied the method of Galileo—had he believed in it.

One deduction Bacon does make. He admits the converse of his theorem. He says that if by any means whatsoever you are able to excite the motion which is heat, "you will beyond all doubt produce heat." Evidently he himself was far "beyond all doubt" for he never tried out this consequence. Let us then do so for him. Suppose that by means of paddles or other arrangements we produce this motion in cold water. Will the water then become hot? Will it boil? By no means! It will not even be perceptibly warm to the touch. Of course one can always object that we do not by this means produce exactly the right motion. What then *is* the right motion? We have only Bacon's description. If we produce a good imitation, why do we not get at least a good approximation to the temperature that the exact motion would have produced?

There is one experiment that would have settled this matter, but it could not have been carried out in Bacon's time, because it requires an air-pump, and such a thing did not then exist. The experiment is that of boiling water at reduced pressure. By removing the air and vapor by means of the pump, water may be made to boil at any temperature we please even down to the freezing point. Here there can be no question about the motion not being exactly right, for the water is actually boiling and being converted into vapor. Yet it is cold.

The motion that Bacon described therefore does *not* constitute heat. If heat is a motion, it must be some entirely different sort. First of all, it must be invisible; for the red-hot iron shows no visible motions. It must therefore be a motion not of parts of "tolerable dimensions," but of particles so small that they cannot be seen. Finally, this motion, whatever it is, must be an adequate cause of all those effects which Bacon took to *be* heat. These are the conclusions that Bacon should have drawn, and could have drawn from the evidence then available, and would have drawn if he had been a really scientific man. Moreover, they are the conclusions that actually were drawn shortly afterwards by scientific men, as soon as they came to consider the motion theory.

What are we to make of Bacon's remark—"It must not be thought that heat generates motion, or motion heat (though in some respects this be true)?" This connection between heat and motion, which Bacon denies, yet admits conditionally, is precisely what the motion theory is devised to explain. If the connection does not exist, then there is no reason for the theory. But Bacon had no intention of explaining anything. He had no use for his "complete affirmation" after he got it—no idea that any theory or any hypothesis could be of use. For him, the end of a research was a final resting place, a philosophical Nirvana, instead of a new starting point, leading to fresh activities and further discoveries. He having settled for all time "by the most certain rules" the question as to the nature of heat, that matter was ended. With all other questions settled in the same way, the seventeenth century would have marked the culmination and end of the sciences, instead of the beginning of a new era.

Finally, Bacon's method is precisely that of pseudo-science. The devotees of such cults invariably insist that their theories must be true, because, forsooth, they fit the facts. Like Bacon, they fail to realize the possibility that other theories may fit the facts equally well. They do not know that to explain is not enough, that all explanations are hypothetical until their factual truth has been established by experience. They have implicit faith in their perfect logic and correct procedure, just as Bacon had in his "certain rules" and system. And when their impeccable theories turn out to be at variance with the facts, it is the latter that suffer, just as Bacon discarded those "instances" that did not fit his conclusion. Bacon was the father of pseudo-science, and the discoverer of its true method.

William Harvey, who was discovering the circulation of the blood while Bacon was dictating the method of science, said of him: "He writes philosophy like a Lord Chancellor." Indeed, Bacon had no idea of the multiplicity of relations, mostly mathematical, that obtain between phenomena. He had much too simple a conception of scientific research. Each instance was summoned but for a single purpose, to testify as to the presence or absence of the required "nature." To each was addressed the one question: "Guilty or not guilty?" And when all the evidence was in, the verdict was rendered.

What is the secret of Bacon's popularity? It seems to be due more than anything else to the pithy and penetrating manner in which he exposed the errors and fallacies of the ordinary mind. As De Morgan put it,[4] he was "eminently the philosopher of *error prevented*, not of progress *facilitated*." We easily discern these defects in other people, while overlooking them in ourselves. That gives us a feeling of superiority which is very pleasant. Bacon has also provided us with a great number of effective aphorisms, such as his "idols," which are convenient to hurl at the other fellow. And that is very satisfying.

---

[4] Augustus De Morgan, *A Budget of Paradoxes* (The Open Court Publishing Company, 1915), Vol. I, p. 79.

CHAPTER IX

THE MOTION AND THE CALORIC THEORIES OF HEAT

LET US now examine the early statements concerning heat and motion of some of those who were really scientific in their ideas. The first of these is Huygens, the founder of the wave or motion theory of light and of radiant heat. He said[1] that flame and fire must contain particles in rapid motion, for they melt and dissolve the most solid substances. He insisted that all parts of physics were branches of mechanics, that is, that heat, light, etc., were all forms of motion.

Descartes attributed the sensation of heat to a motion communicated to the nerves. So did John Locke, and the latter in several places in his *Essay Concerning Human Understanding*, 1690, and in a little work, *Elements of Natural Philosophy*, which is a marvelously simple and popular account of physics written for a friend in 1706, expressed the idea that "Heat is a very brisk agitation of the insensible parts of the object."

The most striking statement, however, of this early period came from Robert Boyle.[2] He declared heat to be a molecular motion, and when it is generated by mechanical means, it is "new" heat. When a blacksmith hammers a nail, it becomes hot. "Yet there appears not anything," said Boyle, "to make it so, save the forcible motion of the hammer, which impresses a vehement and variously determined agitation of the small parts of the iron." He observed that if a nail that had been fully driven into a block of wood, is hammered, it becomes hotter

[1] Huygens, *Traité de la lumière* (Leiden, 1690; Dover, 1962), p. 3.
[2] *The Works of the Honorable Robert Boyle,* Vol. IV, *The Mechanical Origin of Heat and Cold* (1772), p. 249.

than by the same amount of hammering during the driving—
"for, whilst at every blow of the hammer, the nail enters farther
and farther into the wood, the motion that is produced is chiefly
progressive, and is of the whole nail tending one way; whereas,
when that motion is stopped, then the impulse given by the
stroke, being unable either to drive the nail farther on, or
destroy its entireness, must be spent in making a various, vehement and intestine commotion of the parts among themselves,
and in such a one we formerly observed the nature of heat to
consist." This account, written two hundred and fifty years ago,
could hardly be more accurate if written today.

Robert Hooke asked: "What is the cause of fluidness?" and
answered, "This I conceive to be nothing else but a certain
pulse or shake of heat; for heat being nothing else but a very
brisk and vehement agitation of the parts of a body (as I have
elsewhere made probable), the parts of a body are thereby made
so loose from one another, that they easily move away, and become fluid." And while Hooke explained the "fluidness" of a
liquid in this way, Daniel Bernoulli[3] explained the pressure of
a gas by the impacts of its flying molecules on the walls of the
containing vessel.

Leibnitz[4] asked, what becomes of the living force of two inelastic spheres when they collide and thereby come to rest? He
answered: "It is true that the wholes lose it in reference to their
total movement; but it is received by the particles; they being
agitated inwardly by the force of the collision. Thus the loss
ensues only in appearance. The forces are not destroyed, but
dissipated among the minute parts."

It is to be observed that all of these men attributed the motion that is heat to the *minute* particles of a body, not to those
of "tolerable size." The motions are imperceptible and vibratory,
not gross and visible. Hence these motions can be greater in
water that is near the boiling point than when near the freezing
point, even though no difference is visible; and still greater in
red-hot iron. The production of heat by motion and of motion
by heat is *explained*, as the simple conversion of one kind of

[3] *Hydrodynamica* (Strassburg, 1738).
[4] *Fifth letter to Clarke*, cited by J. B. Stallo, *Concepts and Theories of
Modern Physics* (4th ed., New York, D. Appleton and Company, 1900), p. 81.

motion into another, of visible molar motion, into invisible molecular motion, and *vice versa*. The phenomena of heat are thus brought into the realm of mechanics and under the sway of the known laws of mechanics. None of these men advanced his ideas as a demonstrated theory, established by "the most certain rules," but only as a possible explanation, as an hypothesis that could be used, that could be tested. It was a *working hypothesis,* one that opened new roads, not a final pronouncement that marked the end of the trail. Their utterances taken all together form a pretty complete theory that sounds very convincing. Why was it not adopted forthwith? The answer is that as yet there was no decisive experimental evidence, especially of a quantitative nature, in its favor. The motion theory was at this stage, and for long afterwards, a pure speculation—plausible, attractive, and probably true, but a speculation none the less.

On the other hand, the caloric theory explained all the elementary heat phenomena then known in a perfectly satisfactory manner, and quantitatively too. This last was indeed its strong point, for quantitative relations are the soul of science. Even today, the elementary phenomena are most simply and clearly explained by the fluid theory. The flow of heat is most easily understood by analogy with the flow of water, temperature by comparison with water level. Even the motive power of heat, we have seen, was readily explained by Carnot as due to a fall of heat, analogous to a fall of water. In all of these matters the motion theory did not help in the least. Rather it added to the difficulties. First of all, it was impossible to get a quantitative grip on it. How indeed were we to measure a quantity of vibratory motion, and what were we to understand by a flow of motion, or by a fall of motion? These questions do not vex us today, but a century ago they were real difficulties. Secondly, the motion theory was purely an account of the *origin* of heat. It shed no new light on its behavior. Now in studying this behavior, it did not in the least matter whether the heat employed was specially created for the occasion, or was drawn from an existing supply. There was no urgent need of a theory of its origin. And when the origin of a thing is obscure, it is generally safest to assume that it always existed, and to explain everything by redistributions.

In this the calorists were not so unsuccessful as we today, from our vantage point of superior knowledge, are apt to imagine. After some one else has killed the enemy, it is easy enough to kick the corpse. But it was a different matter to tackle him when he was alive and vigorous.

Friction, percussion, and the sudden compression of a gas *apparently* produce heat. But we do not observe this supposed creation. All that we observe, all that our instruments can indicate, is that a *rise in temperature occurs*. That this is due to a creation of heat, is a pure assumption; and scientific caution requires that we avoid an assumption if it is possible to get along without it. This the calorists did. They were able to explain the rise in temperature on the basis of heat already existing.

Every body at ordinary temperatures contains a large and unknown quantity of heat, partly sensible, partly latent. We customarily measure the heat content of a body up from the freezing point of water, because we do not know how much heat was required to raise it from the absolute zero. We do not know the specific heats at low temperatures; we do not know what changes of state may have occurred on the way up. Bodies expand when heated, and this itself is a sort of change of state that undoubtedly absorbs a latent heat, just as do those larger expansions that take place at the boiling and freezing points. Hence a body at ordinary temperatures may contain a large amount of latent heat, that we know nothing of.

The rise in temperature produced when a gas is compressed, the calorists explained as due to a release of latent heat, consequent upon the diminution in volume. Similarly the rise in temperature produced by percussion, they explained as due to the compression of the colliding bodies. And finally, the heat produced by friction they believed due to the compression and attrition of the bodies concerned—the passage from a solid to a pulverized condition being conceived as a "change of state" that released a latent heat. These changes could also be regarded as reducing the thermal capacity of a body, whereby the same heat content produced a higher temperature, some of the contained heat passing from the latent to the sensible form.

The most difficult thing for the calorist to explain was the enormous quantity of heat produced by combustion. Yet he was not daunted even by this problem. A gas, because of its large volume and the fact that it has already passed through the two major changes of state, contains far more latent heat than any liquid or solid at the same temperature. There was no difficulty in supposing oxygen to be particularly rich in this latent caloric, and since most oxides are solid, it must necessarily part with a large quantity in thus passing from a gaseous to a solid state.

But while the caloric theory easily accounted quantitatively for the behavior of heat once produced, it soon began to suffer quantitative strains in accounting for the origin of the heat produced by friction. The first to put a severe strain of this sort upon it was Benjamin Thompson, Count Rumford. The latter in 1798, while superintending the boring of cannon in Munich, was much impressed by the large amount of heat developed during the operations, and proceeded to measure it. For this purpose, he surrounded the metal blank with a wooden box containing two and a half gallons of water, in order by the rise in temperature of the water, to measure the heat liberated. Using a blunt borer in order to increase the friction, he found that as the operation proceeded the temperature rose steadily, until at the end of two and a half hours, the water boiled; and continued to boil so long as the boring continued. This of course would surprise no one today, accustomed as we are to seeing the oil catch fire in a hot box, or the brake lining of an automobile to burn. But at that time, the spectators were enormously astonished to see so large a quantity of water made to boil without the use of fire. Rumford found that the rate at which heat was developed in this experiment, was equal to that of nine wax candles all burning together. And apparently this rate could be kept up indefinitely, so long as the borer turned and any metal remained to bore.

In another experiment Rumford measured the amount of metallic dust that resulted from the boring. A blank weighing 113 pounds was used. After 960 turns, the temperature had risen from 60 to 130 degrees Fahrenheit, and not quite two ounces of metallic dust was produced. "Is it possible," he exclaimed, "that the very considerable quantity of heat produced in this experiment . . . could have been furnished by so inconsiderable a quantity of metallic dust, and this merely in consequence of a change in its capacity for heat?" He also measured the specific heat of the metal before boring, and that of the metallic dust that resulted. He found no difference.

Nevertheless, the calorists were not convinced. It is perfectly possible, they insisted, for the metal to part with a considerable latent heat during the "change of state" from solid mass to powder, and yet show the same or even a higher specific heat than before. Thus when steam condenses to water, it parts with a large latent heat; yet the specific heat of water is twice that of steam. Thermal capacity, measured in the usual way, is no index to heat content, and since we know nothing of the magnitude of the latter, we can set no limit to the amount of heat that a change of state may liberate. It is not at all impossible that the "inconsiderable" quantity of metallic dust may in truth be the source of the *seemingly* inexhaustible supply of heat that Rumford obtained.

Power was furnished for Rumford's experiments by two horses, which tramped around in a weary circle, and turned a sort of capstan. Rumford remarked that the heat was really produced by the work of these animals, and ultimately by the oxidation of the fodder they had consumed, but that more heat could have been developed by burning the fodder directly. In short, there was a considerable loss in this double conversion, of heat into work and back again into heat.

Rumford was thus convinced that work is convertible into heat, and *vice versa,* and that the nature of heat is motion. He said that a sponge when squeezed cannot indefinitely give out water. On the other hand, a bell gives out sound as often as it is struck, and there is no limit to the amount of sound it can ultimately give forth. Water is a substance, sound is motion. Heat is typified, he said, by the vibrating bell, not by the evaporating sponge. He concluded his report to the London Royal Society with the following words: "In reasoning on this subject, we must not forget *that most remarkable circumstance*, that the source of heat generated by friction in these experiments appeared evidently to be *inexhaustible*. It is hardly necessary to add that anything which any *insulated* body or system of bodies can continue to furnish *without limitation* cannot possibly be a *material substance;* and it appears to me to be extremely difficult, if not quite impossible, to form any distinct idea of anything capable of being excited and communicated in these experiments, except it be MOTION."

Rumford also observed that a gun is more heated when fired with a blank charge than with a ball; and explained that in the latter case a part of the heat is *consumed* in propelling the ball.

In the following year Humphry Davy performed some even more convincing if less spectacular experiments. He caused two blocks of ice to rub together, and converted them almost entirely to water by the heat developed by their own friction, although the whole apparatus and the surroundings were kept at the freezing point. Here there was certainly no diminution in thermal capacity, for the specific heat of water is twice that of ice. Furthermore, as was well known, the change from ice to water requires the *absorption* of a latent heat. How then can this operation *both liberate and absorb* a latent heat, as the caloric explanation demands?

In another experiment, Davy placed a clockwork on a block of ice, surrounded also by slabs of ice, and placed the whole in a vacuum. By means of the wound up clockwork, he caused two metals to rub together, and in this way developed enough heat to melt some wax. It is to be noted that this system is completely isolated both thermally and mechanically. There can be no inflow of heat, because the ice barrier must be melted first. There is also no inflow of mechanical energy from the outside. The potential energy of the clockspring is expended in doing work, and the change of state produced is again one that requires the absorption of a latent heat, and therefore cannot possibly also liberate a latent heat.

Despite these telling blows, the caloric theory lingered on for more than half a century, until the final death blow was delivered by Joule in 1843.

To return to the motion theory, we must observe that mere motion does not produce heat. It is arrested motion, resisted motion, that produces heat. A meteor may course about the solar system for ages at any speed whatever without developing any heat. It is only when it encounters the resistance of an atmosphere, or hits a planet, that it bursts into flame. But resisted motion is work. Hence it is the work done, that causes the heat to appear. But this does not yet prove that heat is motion, as Rumford supposed. By expending work in turning a dynamo we produce an electric current, and this in turn by means of the motor will produce mechanical work. Nevertheless electricity is not a form of motion. It is not even a form of energy. There is such a thing as electric power, just as there is water power. In this case, in fact, the analogy is perfect. The dynamo does not create electricity. It merely sets in motion electricity already existing. The motor does not consume electricity. It merely utilizes a fall of electric potential, a drop in electric level. Electricity, in fact, appears to be identical with matter. We have never been able with certainty to separate matter from its electric charge. The neutron is a close combination of a proton and an electron. Hence the mutual convertibility of work and heat does not prove the latter to be a form of motion. The two propositions require separate lines of evidence.

CHAPTER X

THE CONSERVATION OF ENERGY— MAYER, 1842

IN 1840, Julius Robert Mayer, a young physician of Heilbronn, then in his twenty-eighth year, was sent as ship's doctor on a voyage to Java. There, while bleeding patients, he was impressed by the bright red color of the blood taken from their veins. It was much brighter than the color of the blood taken from his patients in Germany. This trivial circumstance, perhaps often before observed but not heeded, started a train of thought in the young physician's head that was destined to make him one of the greatest, but also one of the most unfortunate, discoverers of the nineteenth century.

Mayer knew, from the researches of Lavoisier, that the heat of the body is produced by the oxidation of the blood. In a hot country, there is less need of bodily heat, hence less oxidation, and less difference in color between the arterial and the venous blood. There the matter might have rested. The explanation was complete and satisfactory. But Mayer's thought went on. He asked himself some further questions.

How much heat does the oxidation of the blood have to supply? Obviously it must supply the heat lost by radiation. The body maintains its temperature constantly above that of the surroundings, and therefore is constantly losing heat to the surroundings and thereby warming them. But there is another way in which the body can warm the surroundings, and that is by friction, percussion, or by any sort of work—for the latter always eventually produces heat. What is the origin of *this* heat? Unless we are to suppose, said Mayer, that it is created out of nothing, it must also have been produced by the oxidation of the blood. Thus the latter is called upon to supply both the heat that warms the surroundings directly by radiation from the body, and that which warms them indirectly through the connecting link of mechanical work. And since in both cases the heat produced is proportional to the oxygen consumed, so also must be the mechanical work which temporarily takes the place of one part of it. Heat and work must therefore be *equivalent,* and mutually convertible in some fixed ratio—one unit of the one corresponding to so many units of the other. This is the essence of the conservation of energy.

It is to be noted that Mayer at once assumed that the heat produced by friction, percussion, and eventually by any sort of mechanical work, was *new* heat, which took the place of the mechanical work spent, and not heat somehow squeezed out of the bodies from a supply already contained in them. Doubtless he was not well acquainted with the caloric theories on these subjects, and so was not hindered by them from jumping innocently to this obvious conclusion. His ignorance was thus an asset at first, but it became a considerable handicap later, when he endeavored to make his theories known and understood. He evidently realized that this would be the case, for he applied himself diligently to the study of physics and to the further development of his ideas, becoming so absorbed in this work that he shut himself up in his cabin, and showed none of his shipmates' interest in the remote parts of the world they were visiting. On his return to Heilbronn, he continued his studies with equal absorption, and could scarcely be brought to talk about anything else. As an instance, his friend Rumelin, afterwards Rector of the University of Tübingen where Mayer had studied medicine, relates the following incident:[1]

We were walking one day along the public road discussing these things, when the diligence with four steaming horses passed us. "What in your opinion," asked Mayer, "is the physical effect of the muscular force of those horses?" I replied that I could think of nothing except that the weight of the horses, of the carriage and its contents, had suffered a displacement in space, which without such expenditure of force, would not have occurred. "But," said Mayer, "suppose them to

1 Rumelin, *Reden und Aufsätze* (Tübingen, 1881), p. 350.

pull up halfway and drive back to Heilbronn—what then is the physical effect?" I replied that two displacements in space would then have occurred, the first of which would have been neutralized by the second. Mayer retorted that he could not call this a physical effect. It is quite indifferent, he urged, whether the passengers be landed at Heilbronn or at Oehringen—whether any final displacement occurs or not. The translation of the carriage is the motive and an incidental consequence of the labor of the horses, but not its physical effect. The heating of the horses, the increased oxidation of the food they have consumed, the frictional heat which the moving wheels have left in blue stripes along the road, the consumption of grease at the axles—these are not mere incidents, such as I had held them to be, but the motion of the horses and their mechanical work transform themselves into these heat effects, and indeed in accordance with a fixed quantitative relation, to find and to formulate which, he regarded as his most important task, although he had no longer any doubt about the correctness of the principle.

Another incident is related by Mach.[2] Mayer had gone to consult the physicist Jolly. The latter had some difficulty in understanding him, and finally exclaimed: "But if what you say is true, then water should be warmed by merely shaking it." Mayer left without a word. Several weeks later a man rushed into Jolly's room exclaiming, "And so it is." It was Mayer, whom Jolly scarcely recognized any more, but who seemed to think that Jolly could not possibly have been occupied with anything else in the meantime.

The most obvious way to determine the "fixed quantitative relation" between work and heat would have been to arrange an experiment like that of Rumford, but to measure the work done by the horses as well as the heat produced by the friction. But Mayer had no facilities for elaborate experimentation, no money, but little knowledge of physics, and his duties as town physician consumed most of his time. His only hope, therefore, was to find some existing physical data from which the calculation could be made. How he succeeded in this, we shall presently relate.

In 1841, finding that he had little time for his studies, and to secure himself as he said against "eventualities," he wrote a short account of his theory and sent it to Poggendorff's *Annalen der Physik*. No answer was received, and the paper was not printed. After Poggendorff's death in 1877, the paper was dis-

2 *Wärmelehre* (2nd ed., Leipzig, 1900), p. 246.

covered among his things, and in 1881 Zöllner reproduced it in facsimile in Volume IV of his own works. It was also printed by Weyrauch in 1893.[3]

This first paper of Mayer's contained the essence of the conservation doctrine, but its language was unprofessional and largely incomprehensible. Also it was marred by some grave errors. For instance, Mayer put the energy of a moving body proportional to the velocity instead of to the square of the velocity, an error that was committed by Descartes two centuries earlier, and was corrected by Leibnitz. It stamped Mayer as at least somewhat behind the times. Poggendorff has been much censured for ignoring this paper. But it must be said, that editors and professors are continually inundated by communications from uninformed persons, who imagine they have made great discoveries. Invariably the idea is based on misconceptions. To answer these people at all is to invite interminable controversy. They will neither be taught nor corrected, often resort to abuse, sometimes even to threats. The *Annalen* was the foremost physical journal in Germany. It had a reputation to sustain. Mayer, with his errors and jargon, was doubtless taken for a crank.

In 1842, Mayer wrote another short account of his theories, which he called, "Remarks on the Forces on Inorganic Nature." This he sent to Liebig. It was accepted and printed in the *Annalen der Chemie und Pharmacie*.[4] Mayer had in the meantime corrected his most serious errors, but his language and method were still archaic. It is doubtful if even the liberal-minded Liebig would have accepted the article, had not his own investigations led him to somewhat similar conclusions.

Mayer uses the word *force* where we now use the word *energy*. Since the former term was applied also to a Newtonian force, a very different thing, some ambiguity results. But other writers of the time did the same. Although the term *energy* had been introduced by Thomas Young in 1808, it was not generally adopted until much later. The term *potential energy* was intro-

---

[3] J. Weyrauch, *Robert Mayer, Kleinere Schriften und Briefe* (Stuttgart, 1893).

[4] An English translation is given in a collection of essays by the founders of the conservation theory, edited by E. L. Youmans: *The Correlation and Conservation of Forces* (D. Appleton and Company, 1865).

duced by Rankine in 1853, *kinetic energy*—in place of *vis viva* or living force—by Thomson and Tait in 1867.

The article begins like a metaphysical dissertation. Mayer drags out the old Latin formulas, *causa aequat effectum,* and *ex nihilo nihil fit* (from nothing comes nothing), to prove that work and heat are equivalent and mutually convertible, and the force (energy) they represent is indestructible—a procedure that was very repelling to scientific men. But as one reads on, it becomes clear that Mayer does not rest his case on these formulas, but on empirical evidence, which he presents. He remarks, for instance, that when two metal plates are rubbed together, the heat that appears cannot be due to their compression and the consequent squeezing out of their contained caloric, because water when shaken up is warmed by the internal friction—and *expands.*

In the last paragraph, Mayer comes to his prime purpose, the calculation of the "fixed quantitative relation" between work and heat. He gives 365 kilogram-meters as the work-equivalent of one Calorie, or what we now call the *mechanical equivalent of heat.* How he obtained this figure is not clearly indicated. He remarks in connection with it, that when a gas is compressed, as by the descent of a piston, the heat developed is the equivalent of the work done by the piston. Joule, on reading this passage some years later, concluded that Mayer had calculated the equivalent from the compression of a gas, and so announced in one of his papers. This was not true, and indeed would have been quite impossible, since no data then existed connecting work and the heat of compression. But the notion became current on Joule's authority, that this was Mayer's method, and has not been wholly dispelled yet.

In 1845, Mayer gave a much longer and fuller account of his views in a pamphlet, "Organic Motion in its Connection with Nutrition," which he printed privately in Heilbronn. In this work, he extended his ideas to the whole physical world, including the organic. Mechanical work, chemical action, heat, light, electricity and magnetism, are forms of force, he said, mutually convertible and equivalent; and the totality of these forces in the universe is unalterable. He thus elevated the conservation principle to a universal law. Most important of all, he now gives the details of his calculation of the mechanical equivalent. Let us examine them.

The heat developed when a gas is compressed, or the cold ensuing when it is expanded, had been explained by the calorists, as we have seen, as due to a change in the specific heat of the gas with its change in volume.

As the gas expands, they said, its specific heat increases, so that the same heat content produces a lower temperature. Or, they said, the contained heat becomes more and more converted from the sensible to the latent form. An extraordinary consequence of this view is that if a gas expands to infinity, that is, becomes a vacuum, its specific heat becomes a maximum, and all of its contained heat becomes latent. John Dalton, the father of the atomic weights, actually attempted in 1800 to measure the specific heat of a vacuum.

But Mayer called attention to a little known and long forgotten experiment of Gay-Lussac's, performed in 1807, and published in a journal that had since been discontinued and become extremely rare.[5] Gay-Lussac connected two equal chambers A and B, Figure 15, by means of a tube in the middle of which was a stop-cock. The whole apparatus was well wrapped with heat insulating materials. The chamber A was filled with air, while in B was a vacuum. Thermometers were inserted in each. When the stop-cock was opened, the air rushed from A to B, its volume thereby doubling. According to the caloric theory, the temperature should have dropped, because of the increase in specific heat with the increase in volume. Instead, the thermometer in A fell only slightly, while that in B rose by an equal amount. Both gradually returned to their original readings, so that there was in the end no change in temperature. Neither Gay-Lussac nor any of his colleagues could explain this result, which so flatly contradicted the caloric theory. And so, as Freud says happens with all distasteful things, it was forgotten. For more than thirty years no one heeded the ugly duckling until Mayer found that it was a beautiful swan.

Mayer's explanation was clear and simple. Ordinarily, a gas expands against an opposing pressure, to overcome which it

[5] *Mémoires de la Société d'Arcueil*, Vol. I, p. 180. Reprinted in Mach, *Wärmelehre* (2nd ed., 1900), p. 463.

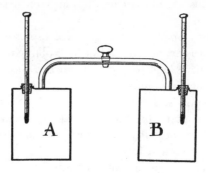

FIG. 15. GAY-LUSSAC'S FREE EXPANSION EXPERIMENT

must do work. This work, said Mayer, is done at the expense of heat contained in the gas, some of which is converted into the work done. The temperature falls, not because of an increase in the specific heat of the gas, but because of an actual decrease in its heat content. Heat is removed by converting it into work just as much as though it had been conducted or radiated away. When the gas expands into a vacuum, there is no opposing pressure, no work to be done, no loss of heat by conversion into work, hence no drop in temperature. The specific heat of a gas therefore does not change with its volume. The temperature changes that occur during compression and expansion are due, as in all other cases, to the actual addition or abstraction of heat. (This statement will require a slight modification, as we shall see later.)

Even the slight temporary changes in temperature that did occur are easily explained by Mayer's theory. In order for the gas to transport itself from the one chamber to the other, it must do a little work. A small amount of its heat is converted into kinetic energy. Hence the slight drop in temperature in A. Arrived at B, the gas swirls and eddies about until its motion is arrested by friction, and its kinetic energy is reconverted into heat. The same amount of heat appears in B as disappeared from A. It has been merely transferred from A to B through the connecting link of mechanical energy. Afterwards, it passes back to A by conduction, and the temperatures in the two vessels become equalized.

Now it had long been known that a gas has two specific heats; that when it is heated under constant pressure, and therefore expands, more heat is required to raise its temperature one degree than when it is heated at constant volume—as in a completely closed container. The calorist explanation was, that since the specific heat of any expanding body increased, part of the heat supplied when the gas expanded, became latent. Hence more was required to raise the temperature one degree than when the gas did not expand. Mayer's explanation was that the extra heat was required to do the work of expansion against the opposing pressure, and was *consumed* in doing it. This heat did not go latent, and continue to exist in some mysterious, dormant, insensible state. It went out of existence—as heat. When the gas cooled and contracted, the external pressure did work on the gas, and this work was reconverted into heat. Hence the "latent" heat is recovered, not because it always existed, but because it is recreated.

Now the heat lost by the expanding gas is the difference between the two specific heats, and is equivalent to the work done. The latter is given by the pressure times the change in volume. All these quantities are known, hence the mechanical equivalent, which is the ratio of the work appearing to the heat disappearing, can be calculated. Mayer proceeded as follows.

Let us consider a cubic centimeter of air. When it is heated from zero degree to 1 degree centigrade, it expands, Mayer said, by 1/274th of its volume (it should be 1/273rd). If we imagine the cubic centimeter to be confined laterally, so that the whole expansion takes place upward, as in Figure 16, its height will be increased by 1/274th of a centimeter. The weight of the atmosphere resting on top of the cube, must be raised by this amount. The weight is 1033 grams (correct). The work required, force times distance, is 1033 times 1/274, equals 3.77 gram-centimeters. The difference between the two specific heats per cubic centimeter for air, according to Mayer's figures, was 0.000103 calorie. This is the heat consumed in doing the above work. Dividing the work by the heat, and changing the units, Mayer obtained 365 kilogram-meters as the equivalent of one Calorie.

This figure was unfortunately considerably in error. Joule obtained in 1849, as the best result of six years of accurate direct

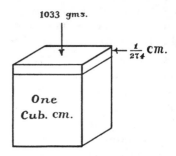

Fig. 16. Expansion of a cubic centimeter of air

measurements, 423.6 kilogram-meters. That Mayer's error was not due to a wrong method but to poor data, especially in regard to the specific heats, was first shown by Tyndall[6] in 1863. Using the newer measurements of Regnault of 1862, and performing exactly the same calculations as Mayer, Tyndall obtained 423.1 kilogram-meters. Clausius[7] by a slightly different calculation obtained 423.8 kilogram-meters. A more perfect agreement between calculation and experiment could not be asked for. In fact, as Clausius remarked, the agreement is a little too good, and is partly accidental.

Mayer used for the specific heat of air at constant pressure, the value given by Delaroche and Bérard, which was 0.267 calorie per gram. The specific heat at constant volume, he calculated from the ratio of the specific heats given by Dulong, which was 1.421. These were the best available data at the time. The modern figures are 0.2412 and 1.402 respectively. From these we may find that the heat consumed by the expansion of one cubic centimeter of air when heated one degree, is 0.0000896 calorie, instead of 0.000103 calorie as found by Mayer. The latter figure is 11½ per cent higher than the former. Raising Mayer's 365 kilogram-meters 11½ per cent, gives 420, which is much nearer the truth. Making slight corrections in his other data, we obtain 423.4, almost exactly the figure obtained by Joule. Nearly all of Mayer's error was thus due to wrong specific heats.

---

[6] *Heat as a Mode of Motion* (6th ed., New York, D. Appleton and Company, 1880), p. 127. [1st ed., 1863.]

[7] *Mechanische Wärmetheorie* (3rd ed., Braunschweig, 1887), p. 56.

This conclusion is confirmed if we apply Mayer's method to other gases, for which the specific heats are still not well known. We then obtain again results which are wide of the mark. The method is therefore not a good one to determine the mechanical equivalent of heat. It requires extraordinary accuracy in the specific heats, and these are particularly difficult to measure with high accuracy.

But while Mayer's method is not practically useful, its theoretical import is great. It gave for the first time the correct explanation of the two specific heats of a gas, and the real reason for the changes in temperature when a gas is expanded or compressed adiabatically. It did away completely with the necessity of assuming enormous changes in the specific heat of a gas with its volume, and emptied the vacuum of its absurd store of latent heat.

The remainder of his pamphlet of 1845, Mayer devoted to consequences and applications of his theory. Most of these are schoolboy commonplaces today; hence we will mention only a few. He explained the heat of fire and of other chemical reactions, as due to the stoppage of the atoms as they rush together under the powerful forces of chemical affinity, just as a meteor bursts into flame when it strikes the earth. He said that the heat which is delivered to the condenser of a steam-engine is less than that which was taken from the boiler, by a portion which is converted into work. This contradicted Carnot, who said that the two quantities were the same, the motive power being developed solely by the "fall of heat" without any diminution in its quantity. In the sun, Mayer saw the ultimate origin of nearly all of our heat and mechanical power. The leaves of plants, he said, capture the sun's rays and by their aid build up the woody fibers, by the subsequent burning of which we regain the heat that was absorbed. Men and animals, by eating the plants, utilize this bottled sunshine. In burning coal or other fossil fuel, we recover the sunshine of a past geologic age.

One of the most important of Mayer's applications of his energy theory was to the explanation of the electrophorus. This is a little apparatus invented by Volta in 1775. It consists of a wax disk to which a negative charge is given by rubbing it with cat's fur. A metal disk provided with a glass handle is then brought near the wax, and receives by induction a positive charge on its under surface and a negative charge on its upper. The latter is removed by grounding. Then, by moving the disk away from the wax, the remaining positive charge may be delivered, say, to a Leyden jar. These operations can be indefinitely repeated, and at each repetition a charge of electricity is obtained nearly equal to the original charge on the wax. Yet the latter is not in the least diminished thereby. A whole battery of Leyden jars may be charged up in this way. What is the explanation of this magical multiplication of a simple charge of electricity? Even the great Volta saw in it something akin to perpetual motion, and it remained a paradox for three-quarters of a century.

Mayer pointed out that to separate the oppositely charged disks against their electrical attraction, requires *more* work, than simply to lift the uncharged disk. Unless we are to suppose that something can be created out of nothing, he said, the electrical effects produced must be due to and equivalent to the *extra* work required to lift the charged disk. We have here a conversion of mechanical work into electrical energy.

In the primitive electrophorus, a great deal more work is done in simply lifting the heavy disk than in separating the charges, and the former work of course produces no electrical effects. To avoid this loss, Wimshurst in 1882 arranged two non-conducting disks to rotate close together on the same axle in opposite directions. On each disk were pasted at intervals small sectors of tin-foil. One set carried the initial charges, the other received the induced charges, both positive and negative, and delivered them to the terminals of the machine. In this way, the only work not converted into electrical effects was that required to overcome the very small friction of the machine. In fact, the machine turns very easily when short-circuited so that the charges cannot build up. But when the terminals are separated and the machine is working, it turns distinctly harder, showing very decidedly the extra work required to produce the electrical energy.

The latter part of this 1845 pamphlet Mayer devoted to applications of his theory to the organic world, a subject which naturally, as a physician, particularly interested him. He made extensive investigations of the food consumed and the work done by men and animals, and even of the oxidation and the work of a single muscle. Meanwhile, he dealt telling blows to the vitalists, and declared all "imponderables" to be relics of Greek mythology.

A second pamphlet, *Contributions to Celestial Dynamics,* Mayer published in 1848.[8] In this he attempted to account scientifically and adequately for the origin of the sun's heat, and his theory was the first to give any promise of adequacy.

The amount of heat which the sun continually pours forth is indeed astounding. According to the measurements of Pouillet, on which Mayer relied, the amount of heat annually received by the earth, would melt a shell of ice all around the globe one hundred feet in thickness. (Modern estimates make it 177 feet.) According to another modern estimate, this heat is 40,000 times the total energy requirements of all living creatures, including the billion mechanical horse-power consumed by the human race. But the earth is only a tiny speck in the firmament of the sun, and receives but one two-billionth of the total heat radiated by the sun in all directions.

Mayer showed that if the sun were a hot body cooling down, it would lose 9000 degrees in 5000 years. If it were a solid lump of coal, its combustion would keep up the temperature only 4600 years. A lump of coal the size of the earth would maintain its fires only five days. And whence, asked Mayer, would we obtain the oxygen for this mighty conflagration? And where, asked Thomson later, could we dump the ashes? Evidently we need a far mightier source of heat than either of these two could provide.

Such a source Mayer believed he had found in the rain of meteors that constantly falls upon the sun, evidence of which we see in shooting stars. The number of meteors that strike into the earth's atmosphere is much greater than is usually supposed. On any clear night, scarcely twenty minutes elapse without a shooting star being seen in some part of the sky, wherever the observer may be situated, and of course just as many fall in the daytime. Occasionally great showers occur. But only an extremely small part of the earth's whole atmosphere comes within the

8 An English translation is to be found in Youmans' collection; see footnote 4, p. 83.

view of any one observer. Mayer estimated the total number of meteors that enter the atmosphere in a year's time to be in the thousands of millions. This estimate is probably correct, but the meteors are exceedingly small, averaging the size of a pea.

The sun is 330,000 times more massive and a million times more bulky than the earth. Hence the number of meteors that fall upon it must be many times greater. Also their energy is far greater. A meteor attracted to the earth from infinity, would strike the latter with a velocity of seven miles per second. Its kinetic energy converted into heat, would produce 17,360 Calories for every kilogram it weighs. A meteor falling upon the sun from infinity, would strike the latter with a velocity of 390 miles per second, and develop 55,000,000 Calories per kilogram. This is 9200 times as much heat as would be developed by the combustion of an equal weight of coal. If the moon should fall into the sun, the impact would provide two years' supply of heat; the fall of the earth would keep the sun glowing 120 years.

Mayer calculated that 200 billion tons of matter would have to fall on the sun every minute to keep up his radiation. Yet so vast is his size that this would add but one three-thousandth of an ounce to each square foot of his surface. It would increase his apparent diameter in the sky by only one second of arc in thirty thousand years, an amount that would be quite unobservable.

Not quite so imperceptible, Mayer had to admit, would be the increase in the sun's mass. This would accelerate the motions of all the planets, thus shortening their periods. The earth's year would be shortened by three-fourths of a second each year. (Modern figures make it 2.8 seconds.) This would be detectable. Since it has not been detected, Mayer concluded that in some ways the radiation of the sun is equivalent to an equal loss of matter, so that his mass remains unchanged. This sounds much like an anticipation of the modern theory of the conversion of matter into energy, but it is not. Mayer had to rescue his theory somehow, and simply pulled down an hypothesis from out of the blue, without any idea of its implications. Incidentally we may remark that the loss of matter required by the modern theory is less than the thousandth part of the gain required by Mayer's theory. The final and decisive objection to the meteoric theory is that if there were anything like the amount of matter that it requires shooting around the solar system, the earth itself would intercept enough of it to become uncomfortably hot. Nearly all of the meteors we actually see are in fact intercepted meteors, and not, as Mayer assumed, bodies drawn to the earth by her own attraction.

These three publications of the years 1842, 1845, and 1848, covering a span of six years, comprise the whole of Mayer's creative work. In 1851, he published a final volume,[9] *The Mechanical Equivalent of Heat,* but it is only a résumé and repetition of his former works. He had by then heard of Joule and acknowledged him as an independent discoverer, but claimed priority for himself. He also adopted Joule's more accurate value of the equivalent, and used it to correct his own calculations. One thing of interest, however, this pamphlet does contain, and that is the story of his Java trip, with which we began this chapter.

The reception of Mayer's work was very different from what he had a right to expect. The great ones ignored him, and the lesser ones ridiculed him. His own townspeople and the local press abused him. Particularly galling was the contemptuous criticism of second-rate teachers and instructors, whose method of refutation was to show that Mayer's ideas were in conflict with the accepted doctrines, and to harp upon his errors and ignorance. And these self-appointed official executioners assumed an air of superiority and of authority that ill sorted with their actual lack of understanding and penetration, but which, unfortunately, mightily impressed the populace—always delighted with a victim which affords them the rate pleasure of a feeling of superiority.

It was of course to be expected that Mayer should encounter some opposition. Every new idea must fight for acceptance, and the newer the harder—and this is proper. Science would be a poor thing if swayed by every wind. But the opposition to Mayer was much like that of a gang trying to prevent an outsider from breaking in. It increased in truculence, the more he seemed likely to succeed.

And while Mayer was struggling for recognition, he had the chagrin of seeing all his discoveries made elsewhere by others and credited to them, and the whole conservation doctrine regarded as a "foreign discovery." In 1843 Joule discovered the

9 English translation in Youmans' collection; see footnote 4, p. 83.

convertibility of work and heat and measured the mechanical equivalent. Many of his statements parallel exactly those of Mayer. In 1845 Holtzmann[10] independently calculated the mechanical equivalent by precisely Mayer's method, and obtained substantially the same result, 374 kilogram-meters. In 1847 Helmholtz independently discovered the conservation of energy and applied it to all branches of physics. In 1853 Waterson[11] independently proposed the meteoric hypothesis, and William Thomson in the following year [12] enthusiastically advocated it and developed it further. There was not a thing that Mayer said or wrote, but some one else said or wrote it, and received applause for it. And when at last Mayer's 1842 article became known, it seemed belatedly to claim credit, on the basis of an a priori deduction, for the discovery that Joule had made experimentally in the following year.

Joule, on reading this article, as we have mentioned, erroneously concluded that Mayer had calculated the mechanical equivalent from the compression of a gas. Since, disregarding the forgotten experiment of Gay-Lussac, it was Joule's own experiments that first demonstrated conclusively that the compression of a gas actually produces new heat, he decided that Mayer's method was based on an unproved assumption and was therefore "illegitimate" and his claim for priority without foundation. Joule can hardly be blamed for his mistake, because Mayer was vague; nor can Mayer be blamed for his vagueness, because he had to compress everything into a few hundred words. But it is not a little surprising that the error was never corrected even after the 1845 pamphlet, in which Mayer sets forth his method fully and clearly, became available in England through Tyndall's translation in 1862. Instead, it was precisely at this time that the controversy (led by P. G. Tait) entered upon its bitterest phases, as though its aims were to suppress Mayer at all costs, and, as Clausius remarked, to claim the whole energy theory for the British nation. It is a curious fact also that Youmans' collection of essays by the founders of the conservation theory, published in 1865, contains Mayer's works of 1842, '48, and '51, but omits the most important of them all, the 1845

---

[10] *Ueber die Wärme und Elasticität der Gase und Dämpfe* (Mannheim).
[11] At the Hull meeting of the British Association.
[12] *Transactions of the Royal Society of Edinburgh.*

pamphlet. One may read this book from end to end, and still not learn precisely what was Mayer's method of calculating the equivalent.

Hence, Joule's assertion that Mayer's method was illegitimate became a fixed dogma, that has permeated the literature and the textbooks more or less ever since. Thus one may read in the Eleventh Edition (1911) of the *Encyclopædia Britannica,* article Mayer, as follows:

> He claims recognition as an independent a priori propounder of the "First Law of Thermodynamics." . . .
> It has been repeatedly claimed for Mayer that he calculated the value of the dynamical equivalent of heat, indirectly, no doubt, but in a manner altogether free from error, and with a result according almost exactly with that obtained by J. P. Joule after years of patient labor in direct experimenting. This claim on Mayer's behalf was first shown to be baseless by W. Thomson (Lord Kelvin) and P. G. Tait in an article on "Energy" published in *Good Words* in 1862, which gave rise to a long but lively discussion. A calm and judicial annihilation of the claim is to be found in a brief article by Sir G. G. Stokes, *Proc. Roy Soc.,* 1871, p. 54. See also Maxwell's *Theory of Heat,* chap. xiii. Mayer entirely ignored the grand fundamental principle laid down by Sadi Carnot—that nothing can be concluded as to the relation between heat and work from an experiment in which the working substance is left at the end of an operation in a different physical state from that in which it was at the commencement.

Yes, and Joule equally ignored "the grand fundamental principle" (these are Tait's words) in every one of the experiments by which he measured the "dynamical equivalent." And it was Thomson's fatal attachment to "the grand fundamental principle" that made him the last of the great physicists to yield up the caloric theory and accept the conservation of energy. In the Twelfth Edition of the Britannica (1929), the article on Mayer is shortened and softened a little, but he is still given no credit as an independent discoverer.

Thus, when Mayer's work finally became known to the "great ones" they ridiculed it no less than the little ones had done, and with far more devastating effect. They ridiculed his metaphysical deduction, his "Dog-Latin Dogmas," his "illegitimate method," his wrong result, and never ceased to hold up against him that he made no experiments. Yet it cannot be denied that a result derived from experimental data already on hand, is just as much

an experimental result, as one for which the data are freshly pro-
cured. And certainly it required far more penetration to unearth
the data needed, and to devise a method by which they could
be used, than simply to measure the work consumed and the
heat produced by friction, with all the facilities of a well-
equipped laboratory, a well-lined purse, and plenty of leisure
time. With none of these advantages, Mayer accomplished his
end, and in five short years placed himself among the foremost
discoverers of his century.

Furthermore, it is not true that Mayer made no experiments.
He did make such simple experiments as he could. In particular,
he performed qualitatively exactly the experiment that Joule
afterwards carried out quantitatively. That is, he shook up
water, and ascertained that a temperature rise occurs. Consider-
ing the extreme smallness of the temperature rise obtainable in
this way, it was no mean accomplishment merely to detect it with
crude apparatus. Try it for yourself. Mayer had no facilities for
elaborate and exact research, and it is nothing to his discredit that
he doubtless lacked also the necessary knowledge and technique.
Finally, neither Mayer nor Joule nor Jolly, as we shall see, was
the first to propose this experiment.

To some extent Mayer's lack of early recognition was his own
fault. He was not to blame of course for the fact that his papers
were printed in obscure places. Reputable journals refused them.
But he was to blame for their misleading titles. Who, for instance, would have dreamed of finding an important physical
discovery in a pamphlet on *Nutrition,* privately printed by an
obscure small-town physician? It was highly improbable. Mayer
seemed to think that scientific men are continually poking around
in obscure corners on the chance of discovering something important, instead of attending to their own researches where the
chances of discovering something are much greater. And from
the accounts of Rumelin and others, and from the Jolly incident,
it appears that he talked too much, and assumed too much that
people had nothing better to do than to listen to him. He was
also too dogmatic and impatient of opposition.

But the effect on Mayer of the treatment he received was dis-
astrous. "Either," he cried out, "my whole method of thought
is anomalous and perverse and then my proper place would be
a mad-house, or I am rewarded with scorn and ridicule for the

discovery of important truths." He grew more and more frantic as the ridicule continued. He could not sleep. One night in May, 1850, lying awake and growing more and more excited, he at last got up and threw himself out of the window to the paved street two stories below. The fall did not kill him, but added physical to his mental sufferings. Finally, he did go mad, and in 1851 was committed to an asylum, where he was harshly treated. Two years later he was released, but never quite recovered his mental balance. The remaining twenty-five years of his life he spent cultivating vineyards, with occasional sojourns in nerve-healing institutions.

Toward the end of his life, however, he received some recognition and acquired some staunch defenders. The first of the "great ones" to recognize him was Helmholtz, who, in a popular lecture, "On the Correlation of the Forces of Nature," delivered in 1854 in the presence of many professional physicists, [13] described the work of Mayer and upheld his priority. Thereafter, whenever Helmholtz had occasion to discuss the new doctrine, he never failed to refer to Mayer as its original propounder.

One unfortunate effect of this recognition, however, was to increase the noise and the harm of the conflict, by bringing the big guns, so to speak, into action. It also ranged England and Germany on opposite sides, and so brought national feeling into play. (Tait even appealed to "scientific patriotism.") One Englishman, however, rallied to Mayer's support. John Tyndall, in 1862, delivered a lecture "On Force" at the Royal Institution, wholly devoted to an account of Mayer's work. [14] The result of this chivalrous act on the part of Tyndall, was to call down upon his own head some of the anathema that had been previously bestowed upon Mayer. In the same year Tyndall delivered at the Royal Institution a series of twelve lectures on "Heat as a Mode of Motion," which were published the following year under the same title. This famous book [15] gave a full and popular account of the new energy theory, before any of it had got into the textbooks. Full justice was done to both Mayer and Joule. There is room in the sky, said Tyndall, for both these luminaries.

13 Helmholtz, *Vorträge und Reden* (Braunschweig, 1865).

14 Printed in *Fragments of Science* (6th ed., New York, D. Appleton and Company, 1897), Vol. I, Chap. XVI.

15 Published by D. Appleton and Company. Later Editions contain additional lectures.

Tyndall relates [16] that while preparing his lecture "On Force," wishing to make himself acquainted with all that Mayer had written, he wrote to two friends in Germany, Helmholtz and Clausius, requesting them to furnish him with a complete list of Mayer's works. Both responded, and Clausius was so good as to procure the books and send them. Before procuring them, Clausius had written to Tyndall saying that he did not believe that anything of importance would be found in them. Before sending them, however, he took occasion to read them. In a second letter, he wrote: "I must here retract the statement in my last letter, that you would not find much matter of importance in Mayer's writings: I am astonished at the multitude of beautiful and correct thoughts which they contain." And he went on to point out the various important subjects in which Mayer had anticipated other eminent writers. Clausius also wrote to Mayer, praising his work, and calling attention to Tyndall's lecture. Mayer responded gratefully to both. [17]

In 1871, The Copley Medal of the Royal Society, its highest honor, was awarded to Mayer, after the same medal had been in the previous year awarded to Joule. Tyndall made addresses on both occasions.[18] In 1878 Mayer died.

The conservation of energy and the motion theory of heat, being now both accepted, have become so interwoven that most people take them to be one and the same thing. But this is not so. They are, as we pointed out at the end of the previous chapter, separate propositions requiring separate lines of evidence. Mayer asserted the equivalence of work and heat, but denied the motion theory. He said in his 1842 article: "Before it can become heat, motion—whether simple or vibratory—must cease to be motion." This was held to be a gross error on his part by his English critics, and to invalidate any claim he might have to the energy theory. For, by the time (1862) that Mayer's works became known in England, the motion theory had become fully accepted and incorporated in the energy theory. But at the

---

[16] *Fragments of Science*, Vol. I, p. 382.

[17] Clausius also reports this incident in his *Mechanische Wärmetheorie*, p. 394, and prints Mayer's letter in response to his own, p. 402 (3rd ed., Braunschweig, 1887).

[18] Both are printed in *Fragments of Science*, Vol. I, Chaps. XIV and XX.

time Mayer wrote (1842), the caloric theory was in full force. The motion theory was not then accepted by a single physicist of note. The evidence for it was not considered conclusive by any of them. In denying it, Mayer simply refused to go beyond the evidence he had, and to assume a hypothesis that was not essential to his own theory. He made a mistake, perhaps, in denying it so positively. He should have taken a more agnostic attitude. But at that time, every one was denying the motion theory with something of that vehemence which one exhibits when the battle is going against him. It is not strange, therefore, that Mayer should have denied it also, and have refused to burden his already unpopular ideas with an official outcast.

But the motion theory was not accepted by every one even in 1862 and after. The atomic theory of Dalton had been denied by Wollaston, Ostwald, Mach, Helm, and others; and with the atoms went overboard, of course, whatever motions had been ascribed to them. Ostwald founded what he called the "Science of Energetics," a science which refused to go beyond immediately verifiable facts and relations, or to admit entities which could never be directly observed. All mechanical models and explanations were regarded as mere pictures, helpful perhaps to minds untrained in abstractions, but corresponding to nothing real. The only theories admitted were mathematical relations between measured quantities. This school applauded Mayer's rejection of the motion theory, and adopted his energy theory. Its aim was a physics free of hypotheses.

This school flourished for a time, but the course of scientific events went against it. The discoveries of the early twentieth century made the atoms so real that they could no longer be denied, and even Ostwald finally had to capitulate. But later there was again a strong trend in the direction of the pure empiricism of the energetic school. The Newtonian mechanics have been superseded. Force has become an unpicturable curvature of space. The too, too solid atoms have dissolved into systems of waves, about which we know nothing but the equations they obey, and even these are haunted by a "principle of Uncertainty." Nothing substantial remains but "pointer readings" and mathematical relations connecting them.

CHAPTER XI

THE MECHANICAL EQUIVALENT
OF HEAT—JOULE, 1843

A VERY different story is that of James Prescott Joule, son of a wealthy brewer of Manchester, and later himself proprietor of a large brewery. As a boy he was delicate and was taught at home until the age of sixteen. He then studied for two years under the chemist John Dalton. This comprised the whole of his schooling. At an early age he showed a great aptitude for mechanics, and a room was fitted up for him in his father's house as a workshop and laboratory.

At this time, the brilliant researches of Faraday had led many people to believe that electricity was about to supplant the power of steam, and it became the dream of the youthful Joule to bring about this transformation in his father's plant. Accordingly, in 1837, at the age of nineteen, he built an electro-magnetic engine operated by a battery. At first he was very enthusiastic about it, and expected quite impossible things of it. But Joule was no ordinary lad who simply played with mechanics. He seems to have had from the beginning a passion for exact measurements. He always looked for the source of an effect. He perhaps had never heard of the "Dog-Latin Dogma" *causa aequat effectum*, but nevertheless he always carefully measured the diminution of the source and compared it with the increase in the effect. This attention to both sides of the equation was characteristic of all his work.

Joule set about to test his electro-magnetic engine. But right at the start, a difficulty presented itself which kept him four years at the task. While it was easy enough to measure the *output* of the engine by the rate at which it would lift a weight, it was not so easy to measure the electrical *input*. It is not a little astonishing that at that time, despite the large amount of electrical research that had been done, no standard units or methods of measuring existed. Every one devised his own units and methods, and expressed his results in arbitrary numbers that had no universal meaning. Joule also had to devise his own methods and make his own instruments. He finally decided to measure the electric current by the weight of water it would decompose in a given time, and later by the rate at which it would deposit a metal in electro-plating. Weighings can always be made with great precision, so that this method is very accurate. It is still the ultimate method by which our convenient direct-reading instruments of today are standardized.

In the course of this investigation, Joule was led to measure the quantity of heat developed by the electric current. This he did by immersing a coil of wire carrying the current in a vessel of water, and measuring the rise in temperature of the water. He found that the heat developed was proportional to the resistance of the wire and to the *square* of the current. This is now known as Joule's law, and is of fundamental importance—but that was little appreciated at the time. Joule prepared a paper "On the Production of Heat by Voltaic Electricity," which he offered to the Royal Society in 1840. It was refused in its complete form, read only in abstract, and later published in the *Proceedings* reduced to twenty printed lines.

Joule now began to realize that the source of the motive power of his electro-magnetic engine was the oxidation of zinc in the battery, and that this was analogous to the burning of coal under the boiler of a steam-engine. By measuring the zinc consumed and comparing it with the work done by his engine, he was able to announce, in a lecture delivered in Manchester in 1841, that by the consumption of a pound of zinc in his battery he was at most able to obtain but one-fifth of the work that the burning of a pound of coal produced in the best Cornish steam-engines. Even if future improvements, he said, should more nearly equalize the performances, the high cost of zinc as compared with that of coal would prevent the magnetic engine from ever becoming useful except for very special purposes. Thus, with his own honest hand, he destroyed the idol of his youth, and perhaps saved many others from pursuing a similar chimera.

The source of the *heat* produced by the electric current, Joule saw, was the heat developed by the oxidation of the zinc. This heats first of all the battery (just as the coal heats the furnace in which it is burnt), but it heats the battery *less* when the current is flowing, than the same amount of zinc dissolved without generating a current. By measurements, Joule determined that the heat developed in the wires plus that produced in the battery, was equal to the heat produced by dissolving the zinc without generating a current. It was as though a part of the heat developed in the battery were transported by the current to the wires and there dissipated.

But an electric current can also be produced by revolving a spool of wire end over end between the poles of a magnet, that is, by a magneto-electric machine—the forerunner of the modern dynamo. A current so generated, Joule found, developed exactly the same amount of heat in the wires as an equal current from a battery. If then the heat is simply transported by the current from its source to the wires, the spool should be correspondingly cooled—for no new heat is generated there. But Joule found that the wires of the spool were also heated by the current traversing them, and to the same extent as by an equal current led to them from a battery. In short, the whole circuit was heated. This heat must therefore be *new* heat, *created* by the current, and not simply transported by it from one place to another.

To make sure of this conclusion, Joule immersed the whole apparatus in a water bath, which served both to insulate it thermally from the surroundings and to enable the heat generated to be measured by the rise in temperature of the water. The only thing then put into the apparatus from the outside, was the work of turning the handle, which in fact is conspicuously greater when the current is generated (because the attraction of the magnet must be overcome) than when the circuit is open and no current is flowing. The heat appearing must therefore be created by this *extra* work, through the intermediate link of the electric current.

The reader will note that Joule's reasoning here is very similar to that of Mayer at Java, and that his explanation of the magneto machine is exactly the same as Mayer's of the electrophorus. But Joule went further. He established the factual truth of his conclusion by measuring, as was his habit, both sides of the equation. The work put into the machine, he measured by causing the handle to be turned by a descending weight suspended by a cord wound up on the axle. He measured the work put in first with and then without the current flowing. The difference was the extra work that produced the heat. The latter he measured by the rise in temperature of the water. He thus made the first direct measurement of the ratio of work to heat—the mechanical equivalent of heat.

By so crude a method, the results of course were not very accurate or concordant. They varied in different experiments from 587 to 1026 foot-pounds per British Thermal Unit (322 to 563 kilogram-meters per Calorie). As the best mean of all, Joule gave 838 foot-pounds (460 kgm.).

But if work can be converted into heat through the connecting link of the electric current, it should be possible to make this conversion directly, and the same ratio should apply. To test this, Joule measured the heat produced by the friction of water, when the latter is forced through small holes in a piston which is pressed down upon it. In this experiment he obtained 770 foot-pounds for the equivalent (422 kgm.).

Both of these experiments Joule described in a paper "On the Calorific Effects of Magneto-Electricity, and on the Mechanical Value of Heat," which he read at the Cork meeting of the British Association in 1843. It attracted no attention whatever.

Despite these discouragements, Joule continued his experiments, varying his methods and seeking ever greater accuracy. He now turned his attention to gases, measuring the work required and the heat produced when a gas is compressed. This was the first time that these two quantities had ever been connected in an experimental research. He obtained in this way 823 foot-pounds for the equivalent. He then measured the heat lost and the work done by a gas when it expands against a pressure, obtaining in this way 798 foot-pounds for the equivalent.

These results are not very concordant, nor are they in remarkable agreement with the previous ones; but roughly they show that the changes in temperature that occur when a gas is suddenly compressed or expanded are proportional to the mechanical work involved, and are not connected with changes in the specific heat. If this is so, then when a gas is expanded into a vacuum, so that no work is done in overcoming an opposing pressure, there should be no fall in temperature. Joule verified this deduction by performing a free expansion experiment similar to that of Gay-Lussac (page 85 above), which was unknown to him. Joule's method was more accurate, however, in that he enclosed the whole apparatus in a water-jacket, whose unchanging temperature showed that no heat was absorbed from the water and went latent in the gas. The experiment proved that there is no large change in the specific heat of a gas on expansion, and confirmed what his measurements had shown, that the fall in temperature when a pressure is overcome, is due to the work done by the gas.

These experiments Joule reported in a paper "On the Changes of Temperature produced by the Rarefaction and Condensation of Air," which was printed in the *Philosophical Magazine* in 1845. No attention was paid to it.

Joule, being now convinced that heat is transformed mechanical work and not a substance, set about to determine the ratio of conversion with all possible accuracy. For this purpose he used a calorimeter in which water, or other liquid, was churned by a set of revolving vanes intermeshing with fixed vanes. The former were driven by descending weights, so that the work put in could be measured. Great care was taken to eliminate all sources of error. Particular attention was paid to the thermometers, for the whole rise in temperature was never greater than half a degree Fahrenheit. Hence they had to be particularly sensitive and accurate. They were especially made for Joule, and could be read to a two-hundredth of a degree. Their testing and standardizing was a whole research in itself. Joule's thermometry was by far the most accurate that had yet been accomplished.

From a large number of these churning experiments, he obtained in 1847 a mean value for the mechanical equivalent of 782 foot-pounds. In that year, he prepared, for the British Association meeting at Oxford, a paper "On the Mechanical Equivalent of Heat, as determined by the Heat evolved by the Friction of Fluids." The chairman of the meeting, because of the great number of papers scheduled, suggested that Joule not read his paper in full, but give a short verbal account of it. This Joule did, and again his report was about to pass unnoticed, when a young man arose, and by his intelligent questioning started a lively discussion. That man was William Thomson, afterwards Lord Kelvin. After the meeting, Thomson had an hour's conversation with Joule, during which, as he afterwards related, he gained ideas that had never entered his head before. On the other hand, he believed that he had also contributed something of value to the conversation by telling Joule about Carnot. It was the beginning of a lifelong friendship.

Although Thomson was much impressed by Joule's paper, he could not bring himself to concur fully, because it conflicted with Carnot's theory, for which he had a great admiration. Joule, however, was well aware of the conflict, for he had discussed it in his 1845 paper, and asserted that the heat delivered to the condenser of an engine was less than that drawn from the boiler by an amount equivalent to the work obtained—the same statement that Mayer had made.

Shortly after this meeting, Thomson, while on a vacation in Switzerland, was walking down the valley of Chamonix, when he saw in the distance a young man coming toward him and carrying in his hand what appeared to be a walking stick, but which he was not using as such. It was Joule, who was carrying a long thermometer with which he proposed to measure the temperature at the top and bottom of a waterfall. In a fall of eight hundred feet, he expected a rise in temperature of one degree Fahrenheit. Thomson joined him, but they did not find a suitable waterfall. [1]

Joule continued his churning experiments with ever more refinements and care. In 1849, he gave as the best and final result of over a hundred such experiments, including some on the friction of metal plates and of mercury, 772 foot-pounds as the work required to raise one pound of water one degree Fahrenheit (equivalent to 423.6 kilogram-meters per Calorie). This figure stood for thirty years.

But the British scientists were still hesitant about accepting Joule's ideas. The smallness of the effects on which he based such large conclusions were looked upon with distrust. They objected that he had nothing but hundredths of a degree to prove his case by. To be a "Joulite" was still to be a little queer.

[1] Sylvanus P. Thompson, *Life of Kelvin* (London, The Macmillan Company, 1910), Vol. I, p. 264.

Thus Joule like Mayer had a struggle for recognition. But, more fortunate than Mayer, he had leisure, money, mechanical ability, and every facility for experimental work. And despite repeated rebuffs he kept doggedly at his work, and persistently brought his results before the highest scientific tribunals, until at last they gained attention if not immediate acceptance.

Thomson was at this time only half a "Joulite." He could not reconcile Joule's convertibility of work and heat with Carnot's theory, which showed so convincingly that motive power is generated solely by the "fall" of heat without any alteration in its quantity. In 1849, Thomson wrote an exposition of Carnot's theory, in which he still adhered to this view. [2] He admitted that Joule's experiments had proved the conversion of work into heat, but he denied that any experiment had as yet shown the contrary conversion to be possible. He demanded more evidence. However, it was not more experiments that finally decided the matter, but the keen theoretical insight of Clausius, who in 1850 showed how to modify the theory of Carnot so as to conform with the results of Joule. This difficulty once overcome, Thomson bcame one of the foremost leaders in the further development of the dynamical theory of heat, which is called *thermodynamics.*

In 1867, the British Association undertook to establish a standard system of electrical units based on mechanical units. Joule made use of the new units to redetermine the mechanical equivalent of heat. He passed an electric current through a coil of wire of known resistance immersed in a water calorimeter. The input he measured electrically, translating it into work by means of the new units. The result came out 782 foot-pounds instead of the 772 he had obtained in 1849. The question arose as to whether the fault lay with Joule or with the new units. Accordingly, Joule in 1878 made a new and most careful remeasurement by the paddle-wheel method, with many improvements and refinements. He obtained 772.55 foot-pounds, substantially the same as his 1849 result. Meanwhile the British Association redetermined their units and found an error in the evaluation of the ohm (the unit of electrical resistance) of more than 1 per cent. This was naturally a great triumph for Joule.

[2] *Carnot's Theory of the Motive Power of Heat, Transactions of the Royal Society of Edinburgh,* Vol. XIV.

Nevertheless, Joule's figure has had to be changed somewhat, for two reasons. Henry Rowland, 1877-79 in Baltimore, repeated Joule's experiment with a huge calorimeter, the paddles of which were driven by a steam-engine. This permitted an indefinite input of energy, and a large rise in temperature. The temperature was carried from 40 to 100 degrees Fahrenheit, a rise of 60 degrees as compared with Joule's ½ degree. Rowland also calculated the equivalent from the rise in temperature from 40 to 50 degrees, again from 41 to 51 degrees, and so on up, obtaining thus some fifty results. It was expected of course that these would agree, except for slight *irregular* variations due to experimental errors. Instead, it was found that they diminished progressively, reaching a minimum near 100 degrees, and then increased. This surprising result was attributed by Rowland to a similar change in the specific heat of water as the temperature rose, and this surmise was later confirmed by direct experiments made by others. Hitherto it had been supposed that the specific heat of water was constant, and Joule of course made that assumption. This was the first circumstance requiring an alteration in his figure. The second circumstance was that Rowland reduced his thermometer readings to the new Kelvin thermodynamic scale (to be described later). Rowland's value for the equivalent was, in round numbers, 778 foot-pounds or 427 kilogram-meters. This figure has stood practically unchanged to the present day.

CHAPTER XII

THE PRINCIPLE OF EXCLUDED PERPETUAL MOTION—HELMHOLTZ, 1847

THE THIRD great independent discoverer of the conservation of energy was Hermann von Helmholtz. He knew nothing of Mayer at the time of his discovery, and only heard of Joule toward the end of his work. In 1847, when only twenty-six years of age, he read before the Berlin Physical Society a paper "On the Conservation of Force." Although this paper was thoroughly professional in character, showed an exact and detailed knowledge of physics, was couched in the proper technical language and fortified with mathematical demonstrations, it attracted little attention, and that little was mostly hostile. It was refused by Poggendorff for the *Annalen,* and was privately printed the same year in pamphlet form. [1]

Helmholtz was thoroughly mechanistic in his viewpoint. He believed as Huygens did that heat, light, electricity, etc., are all forms of motion, more exactly of mechanical energy, and therefore branches of mechanics. They differ only in the nature of the motions, and of the bodies that move. Hence the laws of mechanics already known apply to all, and particularly the law of the conservation of mechanical energy (which we discussed in Chapter VII) is universally true. That was Helmholtz's main thesis. He thus combined the motion theory of heat with the conservation principle, and gave the first comprehensive description of the doctrine in substantially its modern form.

He begins by saying that he prefers to base his theory not on any philosophical principle, but on a purely physical assumption,

[1] Reprinted in Ostwald's *Klassiker der exakten Wissenschaften* (Leipzig, 1902), No. 1.

whose consequences can be developed and compared with experience. This assumption may be either that a perpetual motion is impossible, or that all the phenomena of physics are reducible ultimately to the motions of infinitesimal particles of matter— *mass-points*—under the influence of forces that vary only with the distance. The two assertions, he says, are identical, as he will presently show.

The final aim of theoretical science, he tells us, is to reduce the world to matter, force, and motion; and while this is perhaps not everywhere possible—as for example in the realm of free-will —nature is comprehensible only so far as it is possible. The scientist, whose object it is to make her comprehensible, must proceed on the assumption that she is in part at least comprehensible, and push his researches as far as he can. Helmholtz thus assumes that whatever cannot be mechanically explained, cannot be explained at all.

Matter and force, he says, are inseparable. We know nothing of matter except through the forces it exerts, and nothing of forces except as exerted by bodies. Without force, matter could not affect other matter or our sense organs. Without bodies between which to be exerted, forces could not exist.

All the qualitative differences we distinguish in matter are differences in effects. Science seeks the final unchangeable causes in nature, that is, the *permanent* forces. Substances endowed with permanent forces of a specific nature (unalterable qualities), we call the chemical elements. If we have reduced the world to elementary substances, then the only changes that can occur are spatial, that is, motions. And since the forces are unalterable in time, they can only vary with the distance.

But the force which two extended bodies exert upon each other is the resultant of the forces exerted by all their parts. If the latter have extension, then we must take the parts of those parts, and so on until we arrive at the simple forces between unextended points. All other forces are composites of these.

Pure mechanics and all of its fundamental laws thus reduce ultimately to the motions of mass-points. Helmholtz attempted to prove this of the law of the conservation of mechanical energy. But in a sense it is obvious, for the mathematical equations by which the law is expressed contain only such quantities as forces, masses, distances, velocities. There is nothing about size. In fact, the masses are assumed to be concentrated at points, whose positions are given. Helmholtz therefore attempted to prove what was already assumed. *Why* this assumption is made is another matter, and we shall give some reasons later. And since this law is fundamental to his theory, we shall now give a somewhat fuller description of it than we did in Chapter VII, and a more modern one than Helmholtz gives.

The law states that in any isolated system of bodies moving under the influence of their own mutual forces, the sum of the kinetic energies of all the bodies plus the sum of their potential energies (both always taken at the same moment) is constant. That is, any change in the total kinetic energy is accompanied by an equal and opposite change in the total potential energy. By "isolated" is meant that no energy flows into or out of the system from or to other systems.

The kinetic energy of any particle depends upon its mass and the square of its velocity. Its potential energy depends upon the force to which it is subjected and the distance it can move in obedience to that force. Thus two attracting bodies can move toward each other until they come into contact. The potential energy of such a system of two bodies thus depends upon the distance between them, and is completely exhausted or becomes zero when they are in contact. Two repelling bodies, such as similarly electrified ones, can move apart until separated by an infinite distance. Their potential energy is then zero, and for any other separation depends upon the distance between them. If a system consists of more than two bodies, hence of more than one force, the resultant of the several forces acting on a body must be taken, and its potential energy then depends upon the distance it can move until that resultant becomes zero—that is, until all the forces acting upon it are in equilibrium. At any moment, its potential energy thus depends upon its *position* with respect to all the other force exerting bodies, and the potential energy of the whole system depends upon the relative positions of its parts. Potential energy for this reason is often called *positional energy*.

It follows, that if at any moment we know the positions of all parts of a system, its total potential energy is given. If at a later moment we again know the positions of all the parts, its total potential energy for that moment is also given, and hence also the *change* that has meanwhile occurred. And since the kinetic energy changes equally and oppositely, its change is likewise given. In order then to determine the magnitude of either energy change, we do not need to know the manner in which the system has passed from the one state to the other, nor how long it took. We need only instantaneous photographs, as it were, of the initial and final states. Energy changes are therefore independent of path and time. This is the whole gist and essence of the law.

A simple illustration will make the matter clear. The potential energy of a body is measured by the work it can do in reducing its potential energy to zero. Thus, a ten-pound weight raised ten feet above the ground has a potential energy of one hundred foot-pounds, for this is the work it can do in returning to the ground. Now it has this potential energy at that height no matter how it came to be there, whether we raised it directly by main strength or by a rope and windlass, whether we raised it vertically or pushed it up an incline, whether we took a millionth of a second or a million years to do the job. Conversely, the weight in descending the ten feet will do the same amount of work by whatever path and in whatever time it descends. If its energy is not communicated to other bodies, that is, if there are no resistances, which means that the system *earth–weight* is completely isolated, the body will acquire the same velocity or kinetic energy at the end of the descent, whether it falls vertically, slides down an inclined plane, swings down an arc as a pendulum, or, like a rollercoaster, descends by a wavy path of any form whatsoever. These facts were known to Galileo and Huygens, who embodied them in the fundamental laws of mechanics. The law of the conservation of mechanical energy is merely a generalization of them, extended to include any number of bodies under the action of any mutual forces whatsoever.

Helmholtz after stating the meaning of the law, proceeded to show that it amounts to a denial of perpetual motion. For if we abstract energy from a system, the latter will pass to another state; and in whatever manner and in whatever time it does so, we always get out the same amount of energy. If now we wish to get out this same amount of energy a second time, we must restore the system to the original state. To do so, we must put in the same quantity of energy that we took out, and again the manner and time of doing it are indifferent. If it were otherwise, if there were any way in which we could restore the system with a less expenditure of energy than we took out, then we could use a portion of the latter to make the restoration, and the rest, by repeating the operations, could be multiplied indefinitely. We could thus obtain an unlimited supply of energy from a finite isolated system, and without any permanent change in the latter. The statement therefore that energy changes are independent of path and time, and depend only on the initial and final states of a system, is equivalent to a denial of perpetual motion. This argument, the reader will note, is exactly like that of Carnot, only applied to mechanical energy instead of to heat.

Again, a simple illustration will make the argument clearer. A clock weight in descending does a certain amount of work in overcoming the friction of the wheels, and drives the clock, say eight days. If now we wish to drive the clock another eight days, we must simply wind it up. We must restore the weight to its original height. Now if there were any way of doing this with a less expenditure of work than was derived from its descent, we could arrange that the descending weight raise another equal weight by this more advantageous method, and the remaining work that it can still do would then drive the clock, say one day. But since the operation is indefinitely repeatable, we would thus have a self-winding clock that would drive itself perpetually without drawing power from any outside source.

This is in fact what every mechanical perpetual motion machine attempts to do. The most common form is the over-balanced wheel. This has appeared in numberless varieties, but in all of them a number of weights is arranged about the periphery of a wheel, as in Figure 17, in such a manner that those on the side supposed to descend swing out away from the axle, while those on the ascending side swing in toward the axle. The idea is that the greater leverage of the former will over-balance the lesser leverage of the latter, and so the wheel will be kept eternally revolving. This is equivalent to supposing that the weights will do more work along the path by which they descend, than is required to raise them along the somewhat different path by which they ascend.

Nowadays we deny perpetual motion because it contradicts the law of the conservation of energy. But Helmholtz reversely based

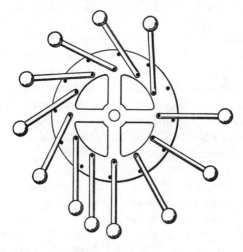

FIG. 17. A PERPETUAL MOTION MACHINE

this law on the denial of perpetual motion. This denial must therefore rest on other and independent grounds. In support of the denial, Helmholtz could only point out that all attempts to produce a perpetual motion had so far failed, that physicists in general were convinced of its impossibility, and that in particular Carnot and Clapeyron had from such denial deduced important laws of heat that were subsequently verified. (He overlooked Stevinus and Galileo, who from such denial derived the basic laws of mechanics.) Later he added as further evidence that as early as 1775 the Paris Academy of Sciences resolved to consider no more claims for perpetual motion, which they put in the same category as circle-squaring. [2] He himself proposed to apply the exclusion principle, or its equivalent, the conservation law, to all branches of physics, and by its means to derive other physical laws, some known, some not yet verified, and thus strengthen the original assumption by showing that it leads to truth. It must be admitted, however, that all these grounds are weak, and Helmholtz undoubtedly relied chiefly on the strong conviction of physicists that perpetual motion is impossible. We have yet to seek then the real source of this strong conviction, and shall do so in a later chapter.

[2] Appendix to *Vorträge und Reden*, Vol. I, p. 404, added in 1883.

The law of the conservation of mechanical energy holds for an isolated system, provided only, as we expressed it in Chapter VII, no complications occur. One of these complications, we mentioned, was a collision of inelastic bodies. Another is a close approach of large bodies under strong gravitational forces by which tidal distortions are produced. In either case, some or all of the mechanical energy disappears, and was formerly supposed to be utterly destroyed. For this reason, the law applies perfectly only to mass-points. Points cannot collide, for they have no size; neither can they be distorted, for they have no shape.

In practice, however, we have to deal with bodies of finite size. The law applies to these so far as they can be treated as simple, indivisible, undeformable wholes, that is, so far as whatever actions take place *within* them do not appreciably affect the actions that take place *between* them. A necessary condition for this is that all forces shall be directed toward or from the centers of mass of the bodies, that is, that they shall be *central* or *Newtonian* forces. Otherwise they will produce rotations or distortions which absorb energy, which take it *out of the system* and put it *into the bodies* that compose the system.

Thus the law holds for the solar system and for celestial motions in general, so long as the bodies do not collide or approach one another so closely that tidal distortions occur. Under these circumstances a planet or a star may be a mass-point as well as an atom or a molecule. The law holds, as Helmholtz further enumerates, for rigid frictionless machines, and for unresisted motions in general. It holds for wave motion in a perfectly elastic medium, for the hydrodynamics of a frictionless incompressible fluid. It holds also for the collisions of perfectly elastic spheres, for these when deformed spring back into shape with no loss of energy. Also, since the force of impact is always directed toward the center of the sphere, no rotations are produced. This is not true of other shapes. (The whole kinetic gas theory is based on these principles.)

But these are all ideal cases, and so are mass-points, one may object, so that the law never applies exactly to any actual bodies. But this is the case with every physico-mathematical law. We cannot deal mathematically or even logically with the things themselves, but only with concepts that fit them more or less closely. A law is not considered to fail if its deviation from actuality is due solely to the imperfect fit of its concepts. No physical law is absolutely exact in its *application*.

The law of mechanical energy, however, is not merely inexact; it breaks down completely in many cases, which Helmholtz proceeds to enumerate. It fails whenever any kind of a resistance is concerned, for this constitutes a system of innumerable minute external bodies which absorb the energy of the system of gross bodies. The latter hence is not isolated. But energy may escape even from a system that is completely isolated from external systems, and that is, as we have mentioned, by passing into the bodies of which it is composed, as when the latter are set into rotation. When inelastic bodies collide, or are distorted by tidal action, heat appears as the result of internal friction. If heat is a form of mechanical energy, Helmholtz says, then the energy of the system is not destroyed, but again passes into the bodies of which it is composed. In a sense, the system is also not isolated. Energy indeed does not escape to an external system, but it escapes to an internal one, to a system within the system. The law is rescued, if for the gross bodies we substitute the minute ones as the individuals of the system considered.

The process does not necessarily end here. If the atoms that compose the molecules are capable of motions among themselves without destroying the identity of the molecules, then energy may escape from the molecules to the atoms. And if the latter are composed of parts, energy may further escape to these parts, and so on. Obviously the possibilities of internal escape are not exhausted until we arrive at the mass-points. These alone have no insides.

That heat is not a substance but motion, that it can be created and destroyed, is thus an essential part of Helmholtz's doctrine, and he supports it with arguments and evidence. First, he shows the insufficiency of the caloric theory to account quantitatively for the heat produced by friction and percussion, on the assumption that it is squeezed out of the bodies concerned by their compression. He points out, that, aside from the fact that this explanation attributes an enormous effect to a frequently immeasurably small condensation, it fails completely to account for the heat produced by the friction of liquids, since in these cases no condensation occurs, but instead, an expansion. This is the same argument that was used by Mayer. Helmholtz exposes the inadequacy of the caloric theory also in other ways, particularly to account for the enormous heat of combustion. He cites the experiments of Davy, and finally thosè of Joule.

That heat can be created, he says, is proved by the heat produced by electricity when the latter is generated by mechanical means. This can be done in two ways. We can produce either static or current electricity. The former can be done by means of the electrophorus, the action of which he explains in exactly the same way that Mayer had done. The electricity accumulated in the Leyden jars will by their discharge produce heat, and there is no possible source of this heat, other than the extra work required to separate the electric charges. For the production of current electricity by mechanical means, he refers to the experiments of Joule with the magneto machine, which showed that the heat produced in this case is similarly due to the extra work required to turn the machine against the magnetic attractions.

These two experiments, Helmholtz concludes, show that heat can be created by mechanical forces, that the phenomena of heat cannot be derived from a material substance that by its *mere presence* produces them, but only from the variations, from the motions, either of a special substance, or of the already known ponderable and imponderable substances, such as for example the electric fluids or the luminiferous ether. Helmholtz thus does not commit himself as to *what* is moved in the case of heat. Nor does he reject *all* imponderables as Mayer did so vehemently.

As evidence that heat can be destroyed, Helmholtz says that he can point only to one experiment, that of Joule on free expansion, which showed that when a gas expands against opposition some of its heat content is consumed in doing work.

Concerning the value of the equivalent, Helmholtz refers to the experiments of Joule. The latter had not at that time given a final result, and so Helmholtz quotes several of his earlier values which ranged from 464 to 521 kilogram-meters (847 to 950 foot-pounds). Helmholtz also cites the calculation of Holtzmann [3] of 1845, which, we have seen, was made by the same method that Mayer had employed, and gave very nearly the same result—374 kilogram-meters. Helmholtz objects that this method

[3] *Ueber die Wärme und Elasticität der Gase und Dämpfe* (Mannheim).

is permissible only after a free expansion experiment has shown that the heat lost by an expanding gas is really consumed in doing work. This is the same objection that was later raised by the Joulites against Mayer. It should apply more strongly to Holtzmann, because the latter never knew and never mentioned the Gay-Lussac experiment, by which Mayer justified his own procedure. However, Helmholtz gives little weight to the objection, saying that the correctness of the method is demonstrated by the harmony of its consequences with experience. Evidently he gave also some weight to Holtzmann's figure, for he remarked that all of Joule's were probably too high—as indeed all of those quoted were.

The two forms of heat, latent and sensible, Helmholtz identified with the two forms of energy, potential and kinetic. When heat enters a body, it produces in general two effects, an expansion and a rise in temperature. To separate the molecules of a body against their attractive forces requires work, just as it does to separate a body from the earth. In either case the potential energy of the system is increased. If the expanding body has also to overcome an opposing pressure, additional work has to be done. These two jobs, which we now call the internal and the external work respectively, consume that portion of the heat which disappears and is said to become latent. The remainder of the heat communicated increases the kinetic energy of the molecules, that is, raises the temperature of the body. Conversely, when heat is abstracted from a body and it contracts, the closer approach of the molecules means an increase in kinetic energy. This corresponds to the apparent reconversion of the latent into sensible heat.

A particularly large amount of heat is required to convert a liquid into a vapor; first, because of the heavy work required to tear the molecules apart; and second, because of the large expansion that occurs. Modern figures show that the former work absorbs by far the greater part of the latent heat. For example, the latent heat of steam at atmospheric pressure is 970 therms, and although water in becoming steam expands to 1600 times its original volume, the work required for this great expansion absorbs only 72 therms. The remaining 898 therms are required to tear the molecules apart. Put in mechanical units, the external work required to evaporate one pound of water at atmospheric pressure is 56,000 foot-pounds; the internal work required is 700,000 foot-pounds. An ordinary kettle will easily boil off a pint of water in twenty minutes, and is then doing the work of a one-horse-power engine.

Helmholtz explained the heat developed by chemical action, such as combustion, in the same way Mayer had done, as due to the kinetic energy developed by the atoms as they rush together in obedience to the powerful forces of chemical affinity. The remainder of his pamphlet is devoted to applications of the energy principle to other branches of physics, particularly to electricity and magnetism. Helmholtz shows that in every case the conservation principle enables us to set up an equation which demands that, whatever else happens, both sides of the energy account must balance. This may seem to be merely a regulatory and error-preventing principle, but it is more. As mathematicians well know, to solve a problem containing several unknowns, we must have as many independent relations or equations as there are unknowns. Otherwise *none* of the latter can be found. Oftentimes the energy equation provides just the one additional relation required to determine *all* the unknowns.

Applying it in this way, Helmholtz derived many laws that previously had been known only empirically, and thus gave them a theoretical foundation. He showed that they were not isolated facts, but parts of a comprehensive system. Because of this unifying power of the conservation law, it has been called the greatest generalization of physical science. But Helmholtz did more. He derived by means of it laws that were not previously known, but have since been verified. He thus made it a tool of research. One example will suffice. He showed that the discharge of a Leyden jar should be oscillatory, for the potential energy of the charges is converted into the kinetic energy of the current. But, as with a pendulum, it overshoots the mark, and the kinetic energy of the current is reconverted into the potential energy of charges of opposite sign. This again is reconverted into the kinetic energy of a reversed current, and so on. Gradually the oscillations die out as their energy is absorbed by the resistance of the circuit, and thus converted into heat, and the more rapidly the greater is the resistance—just as the oscillations of a pendulum are damped by the resistance of the air, and their energy finally converted into heat, and the more rapidly the greater the resistance. The mathematics of the process were worked out by Thomson in 1853; the oscillations were discovered experimentally by Fedderson in 1859; they were applied by Hertz in 1887 to produce electric waves, and are now used for that purpose in radio and wireless telegraphy.

Helmholtz concluded his paper by saying that he believed he had shown that the conservation law is not contradicted by any known facts, but is on the contrary strikingly confirmed by many of them; that he had developed new laws by means of it that must await confirmation by experiment. The object of the paper, he said, which may excuse its partly hypothetical character, was to acquaint physicists as completely as possible with the theoretical importance of the law, and with its practical usefulness as a tool of research. The full confirmation of it should be, he believed, the chief task of the physics of the future.

Unfortunately nobody saw it in that light. A small error in his mathematics at the outset seems to have deafened his hearers to any truth that might be contained in the paper. After all, this was not a physicist to whom they were listening, but a young Prussian army surgeon from Potsdam, who was striking out into a new field. Who was he to demand that future physics concern itself chiefly with his ideas? Yet, before long, that was precisely what future physics did.

Helmholtz had always been intensely interested in physics and had shown ability in mathematics. He wished to pursue a career of research in these subjects, but the limited means of his family required, as the price of a university training, that he study medicine. He did so, but along with it pursued diligently his favorite subjects. For twenty-two years (1849-1871) he taught physiology at various universities. At last in 1871 he was appointed professor of physics at the University of Berlin, and so achieved his ambition of becoming a professional physicist. He died in 1894 at the age of seventy-three, after a scientific career of over fifty years, during which he made many important and some epoch-making contributions to physics, to physiology, and to the unexplored borderland between them.

Helmholtz like Mayer propounded a theory of the sun's heat based on the conservation of energy. This he described in a lecture at Königsberg in 1854, "On the Correlation of the Forces of

Nature." [4] This was the same occasion on which he called attention to Mayer.

Helmholtz's theory was that the sun maintains its radiation by shrinking, as a ball of gas must do while losing heat. The falling in of every part of the sun's surface toward the center, means a reduction of potential energy and a consequent increase of kinetic energy or sensible heat. Helmholtz calculated that a shrinkage of the sun's radius of only 100 feet per year would account for its present rate of radiation. This contraction theory of solar energy has today given way to the theory of conversion of matter into energy; the newer theory is necessary to account for the sun's constancy over long geological history, as demonstrated by billion-year-old fossil finds.

Working backwards in time, and assuming with Kant and Laplace that the sun was once a vast nebula extending beyond the orbit of the outermost planet, Helmholtz calculated that in shrinking from that size, the sun must have developed twenty-two million times the amount of heat that he now annually radiates. The shrinkage must have taken more than twenty-two million years, however, for the nebula was doubtless much cooler at first and radiated more slowly. As it contracted it grew hotter, and, according to this theory, the sun must continue to grow hotter until it begins to liquefy, or becomes so dense that it no longer acts like a perfect gas. Once liquid, it will cool rapidly, and in a few thousand years cease to glow altogether.

This was a much more powerful and also more plausible theory than Mayer's. It required no gratuitous hypothesis, like the latter's vast swarm of meteors that would disturb the whole solar system. It depended only on the nebular hypothesis, then fully believed to be true. In fact, it was a necessary consequence of that hypothesis. It gave the sun a lease of life which then seemed enormous and entirely sufficient. The theory was accepted for half a century, and presented in all the textbooks as the true one. But the demands of biological, geological, and finally, of stellar evolution for a much greater length of time, and doubts about the nebular hypothesis, at last undermined it. Fortunately, the discovery of radioactivity and of subatomic

---

4 *Vorträge und Reden* (5th ed., Braunschweig, 1903), Vol. II, p. 80. Helmholtz again describes his theory in his lecture, "On the Origin of the Solar System," 1871, *ibid.*, Vol. II, p. 79.

energy, as possible sources of stellar heat, have made possible an enormous magnification of our time scales, the substitution of billions for millions of years in the life of a star.

CHAPTER XIII

MAYER AND JOULE

IN A LETTER to P. G. Tait, published by the latter in the preface to his *Sketch of Thermodynamics* (Edinburgh, 1868), Helmholtz expressed himself as follows:

> With regard to Robert Mayer, I can well understand the attitude you have taken toward him, but cannot allow this opportunity to pass without saying that I am not entirely of the same opinion. The progress of science depends thereon that new inductions are continually made from the available facts, and that the consequences of these inductions are then, so far as they point to new facts, compared with the reality by means of experiment. There can be no doubt as to the necessity of this second step. It often costs a great expenditure of labor and ingenuity, and he who carries it through well renders a high service. Yet the fame of discovery still belongs to him who found the new idea. The subsequent experimental verification is a much more mechanical sort of work. We cannot unconditionally demand that the originator of the idea be obliged to carry out also the second part of the work. In that case we would have to reject the greater part of the work of all mathematical physicists. Even William Thomson made a whole series of theoretical investigations concerning Carnot's law and its consequences, before he made a single experiment in connection with them. Yet it would never occur to any of us to value those investigations lightly on that account.

This letter brings up the whole question of the relation of theoretical to practical work in science. Faraday had laid down the rule that the credit of a discovery belongs to him who first *establishes* it. His slogan was—work, write, publish. This rule is based on the conception that the object of scientific research is to add to our stock of *ascertained* knowledge, not to our fund of speculative ideas. Mere guesses are thus ruled out, even if

122

they turn out afterwards to be correct. Indeed, if we were to account all those as discoverers whose guesses turned out to be right, we would have to go back to the Greeks and Egyptians for the first vague beginnings of everything we now know, and questions of priority could never be settled. We may account them anticipators but not discoverers.

But when is a discovery established? Faraday was essentially an experimenter. He had no talent for mathematics, although he recognized its value. Both he and his immediate followers laid the chief emphasis on *experimental* verification. But experiments are never exact. Experimental verification, however striking and complete, never establishes anything with absolute certainty, nor for all time. If it did, we could not in this century, after all these years of seemingly thorough and complete verification, have had doubts about the wave theory of light, or about the exactness of the Newtonian law of gravitation. Nor could Charles Lane Poor and Dayton C. Miller with any justification have disputed the Einstein theories.

Furthermore, there are certain laws so fundamental or general that they cannot be established by any single or even by several crucial experiments, but depend for their final acceptance upon their applicability to a large field, or perhaps to the whole field of physical phenomena. Such are the law of inertia, and the law of the conservation of energy itself in its universal sense. There are also theories that concern entities that are not directly observable, like the atoms, the electrons, and the ether. These must always remain more or less speculative, and their acceptance or rejection depends largely upon the world view of scientists at the time. Thus the atomic theory after a long struggle has won—for the present. The ether, the existence of which was at one time regarded as certain as that of coal or iron, is now in disfavor. These theories are, however, *useful* speculations. They are founded on facts; applicable to facts; explain, clarify, and unify large fields of facts. They are not merely fanciful. Hence it turns out that the rule of Faraday, though simple to state, is not so simple to apply.

On the other hand, it is possible to establish certain laws and relations without the help of special experiments. Logical and mathematical deductions from known facts or from well established theories are as secure, or nearly so, as the foundations on which they are based, without further verification. This is the sort of work the theoretical or mathematical physicist does, to which Helmholtz referred. It is just as sound and legitimate a way to discover and to *establish* new truth as the experimental method. Deductions from a hypothesis must be verified, because the hypothesis is uncertain. This process, since Galileo, has so come to be regarded as *the* method of science, and by many as the *only* method, that the older *rational* method has been largely overlooked. Yet an immense amount of important truth has been discovered and established by the latter method—all of the "rational mechanics" down to the time of Galileo, and much else since that time, for it is by no means an abandoned method. If it is not sound, then all this vast body of knowledge must be rejected—except in so far as it has been confirmed by later experience—and its proposers denied the honor of discovery, because they used an "illegitimate" method. That would be ridiculous indeed.

Deductions from a highly evident truth—that is, from one that is an epitome of wide or general experience—are often far more certain than the results of direct experiments. They rest, so to speak, on the general experience of the whole race, rather than on the special and limited experience of a few individuals. Consider, for example, Archimedes' demonstration of the law of the lever. [1] He might have *assumed* the law (that for equilibrium the weights must be inversely as the distances from the fulcrum) as an hypothesis, and have *verified* it by weighing the weights and measuring the distances. But in this way he could have established the law only approximately, and in his day the approximation would have been rough indeed. Besides, he would not have escaped assumptions. How indeed was he to measure his weights? Unless he was to judge them by the feeling, he could only have used an equal arm balance. But how was he to know that when weights equally distant from the fulcrum balance, they are equal. "Oh, everybody knows that!" you will exclaim. Precisely! And everybody knows too that if one of the weights is increased, or moved further from the fulcrum, that side of the balance will descend. From these facts which everybody knows, Archimedes *deduced* the law of the lever. He thereby avoided

[1] Mach, *Die Mechanik* (Leipzig, 1883). Mott-Smith, *Principles of Mechanics Simply Explained* (New York, Dover, 1963), p. 30.

the uncertainties and inaccuracies of direct measurements, and gave the law a much higher degree of certainty than any such measurements could have given, or can give it even today.

All the simpler and more fundamental laws of mechanics have been discovered and *established* in just this way. This is true even of the law of gravitation. Newton expressly declared that he framed no hypotheses, but deduced everything from the most certain principles. He was certain of his system before any verification of it. He expected nature as a matter of course to conform—and she did. [2]

After all, Galileo's experimental method is only a modification of the earlier rational one. The latter seeks to find the most certain principles from which to deduce equally certain results. Galileo showed the possibility of starting with a hypothesis that is not entirely certain, but which can be made so by testing its consequences. But no principles are ever so certain that subsequent verification of the deductions can be entirely dispensed with. Nor, despite the most perfect logic, are the deductions ever quite so certain as the principles. We cannot reason directly from *things*, but only from *concepts*, by which the things are replaced, and which never fit them perfectly. Assumptions have to be made at the start, and usually further assumptions as the reasoning proceeds. We cannot, in short, get rid of hypothetical elements entirely. For these reasons, we cannot safely carry our deductions too far without checking them by comparison with experience. A rational physical theory is like a cantilever bridge that we start to build at one end. However firm the rock may be from which it springs, we cannot build out too far without putting a pier down to earth.

On the other hand, no hypothesis is ever pulled down from the blue. It is already based on observations, and framed in accordance with scientific principles. It already possesses a high probability. If it did not, no scientist would waste his time and labor in testing it. No one, for example, has ever tested the idea that the moon is made of green cheese.

Hence the two methods are essentially one. They differ only in the degree of assurance pertaining to the major premise, and thus in the amount of experimental verification the deductions

2 *Principles of Mechanics Simply Explained,* p. 55.

require.[3] Neither method can establish anything with complete certainty, nor for all time. It follows that one may be a discoverer though he never perform an experiment.

The further question arises as to how far one may be accounted a discoverer, whose theory involves a hypothesis that really needs verification, but which he himself does not provide. The general practice has been, if the theory is later verified, to divide the honors between the originator and the verifier, as near as may be in proportion to the labors and ingenuity of each. This requires judgment and leaves room for differences of opinion. Halley, for example, who predicted the return of the comet of 1682, is given full credit as the discoverer of the first periodic comet. Those who verified the prediction seventy-five years later, had only to watch the skies. Leverrier and Adams, who in 1846, from the perturbations of the planet Uranus, predicted the existence of Neptune, are accounted the discoverers of that planet, although Galle in Berlin was the first human being ever to see the planet. But Galle merely pointed his telescope as directed by Leverrier. Similarly Percival Lowell predicted Pluto. But in this case the Flagstaff Observatory receives a good share of the honors, because the search was long and difficult, required great improvements in technique, and the planet turned out to be much smaller than expected. Maxwell in 1864 developed his electromagnetic theory of light, and was honored even during his lifetime for this remarkable work, although it was not verified previously until after his death, and not all of the evidence previously available was favorable. Hertz is highly honored for his discovery in 1887 of the electric waves predicted by Maxwell, for this required the invention of the Hertz oscillator, of a suitable detector, and the overcoming of many other difficulties in an elaborate two-year research. Lebedeff and Nichols and Hull are honored for their detection in 1900 of the pressure of light, also predicted by Maxwell. Another striking case is that of Einstein, who in 1905, as the result of a bold hypothesis, proposed his now famous photo-electric equation, the basis of television. At the time, it seemed to have no real foundation, either theoretical or experimental, and any experimental test seemed quite hopeless. Nevertheless, when

3 *Ibid.*, p. 56.

eleven years later the equation was verified by Millikan, [4] Einstein was awarded the Nobel Prize for this discovery.

But an unverified theory, to count as a discovery, must be no mere partial glimpse of the truth. It must be a developed theory. It must be based on facts and applicable to facts. It must already have been compared with all available pertinent facts, and seriously contradicted by none of them. It must explain, and explain better than existing theories—perhaps cover a wider field. It must be verifiable. It must lead to consequences by which it can be tested. If the proponent has developed some of these consequences, as Einstein did in the case of his gravitational theory, so much the better; but this is not necessary. Finally, it must in the end *be* verified.

If we apply these principles to Mayer, we must, I think, accord to him priority as discoverer of the conservation of energy in its universal form, and especially in its application to living things. It was in his hands a complete theory. He drew its consequences and made innumerable applications. It accorded with all known phenomena and was contradicted by none. Particularly it explained in a simple unstrained manner and by means of a single formula, all those things which the caloric theory had been able to account for only with difficulty and with the aid of auxiliary hypotheses that bordered at times on the absurd. And it explained much more that was quite beyond the reach of the caloric theory, giving even for the first time a theoretically adequate account of the sun's heat. That Mayer's energy theory turned out to be correct proves that he was a real discoverer of new truth, and should be honored as such.

It must be admitted, however, that Mayer's whole theory was based on a pure assumption, namely, that heat and work are mutually convertible and equivalent. The making of this assumption was the very first act of his thought, as we pointed out at the beginning of Chapter X. Afterwards he supported it by Gay-Lussac's experiment and by shaking up water. These may be regarded as evidence, but not as proof of the proposition. The latter cannot be said to have been really established until Joule's quantitative measurements were made. Even then, as we have seen, it was still possible for Thomson to doubt and to demand

[4] *The Electron* (Chicago, University of Chicago Press, 1917), p. 224.

more evidence. Mayer's theory was thus based on a hypothesis not yet fully established. Nevertheless, he made it extremely probable by the evidence he presented, and highly attractive by the use he made of it. It explained everything so simply and naturally that the caloric theories looked sick and ridiculous in comparison. It united all branches of Physics, bringing even the animate world into the fold. It cannot be admitted that Mayer had no right to use this hypothesis before it was fully established, and see what he could do with it, especially when he did so much that could not have been done with the existing theories. It is the function of theory to go ahead of experiment, and point out the possible paths of progress.

It must be admitted also that Mayer's calculation of the equivalent was based on a *new interpretation* of the existing data, in accordance with his fundamental assumption. Even if he had obtained the correct figure, it would still have needed verification by direct measurement. How otherwise could one know that either the figure or the assumption was correct? But after Mayer's method turned out to be correct, to declare that his brain-child though sound was "illegitimate" was something of a misapplication of mid-Victorian morals.

People were of course justified in doubting the factual truth of Mayer's theory, but not in dogmatically denying it, much less in condemning and ridiculing it. Rather it merited attention, and should have been at once put to test by those competent to do so.

On the other hand, Mayer was not justified in dogmatically asserting it. If we are to criticize him for anything, it must be for not sufficiently appreciating the hypothetical nature of his theory. We cannot blame him for not making experiments that he was unable to make, but we can and do blame him for not realizing that they were necessary. It is possible that Mayer would not have been so dogmatic if he had had the entrée into the higher physical circles, received sympathetic attention and good advice. But his acquaintance was mostly of persons having little or no knowledge of physics. Prejudice and intolerance could only be met with positive assertion. And so Mayer tried to put over his system mainly on the strength of his own conviction, bolstered by a pseudo-metaphysical deduction. That was not very scientific.

We may contrast with this the procedure of Helmholtz. Although the latter's principle of excluded perpetual motion was a far more convincing foundation than Mayer's not yet fully established assumption of the equivalence of work and heat, and although Helmholtz could already avail himself of Joule's work, nevertheless he recognized and even apologized for the hypothetical part of his system, and called upon physicists of the future to test it. He may, like Newton, have had no doubt as to what the verdict of experience would be, but he recognized the need of that verdict. That was the truly scientific way.

E. Dühring, who two years after Mayer's death wrote a glowing account of the latter's life and work, called him the Galileo of the nineteenth century. [5] A less appropriate simile could scarcely be imagined—comparing Mayer with the experimental Galileo! Mayer, with respect to his method, is far more to be compared to Copernicus. The latter too made almost no observations of his own, is even reputed never to have seen the planet Mercury, and certainly never made any observations on that planet. He used data already accumulated, much of it going back even to Ptolemy. In his day, the Ptolemaic system was already in decay. It was being continually patched up with new epicycles and eccentrics, which had to be added for each new discovery—much as the caloric theory required its auxiliary hypotheses. Copernicus was not the only one who felt that more radical reforms were needed. His system, when first promulgated, had nothing but its simplicity and unity to recommend it. It accounted for the planetary motions, not better, but only with fewer devices than the Ptolemaic. And there were objections to it, which neither Copernicus nor any one else at the time could answer. One real advantage it had, however. If true, it provided a means of calculating the relative distances of the planets *from data already on hand,* which the Ptolemaic system could not do—just as Mayer's system provided a means of calculating the temperature changes that occur when a gas is expanded or compressed, which the caloric system could not do. Finally, the original Copernican system, with its uniform circular motions and its retention of a good many epicycles and other devices, was very different from the modern solar system, just as Mayer's original

[5] *Robert Mayer, der Galilei des neunzehnten Jahrhunderts* (Chemnitz, 1880).

energy theory was different from the modern dynamical theory.

But Copernicus required his Galileo, and Mayer required Joule. Even if Mayer's system had become immediately known and recognized, Joule would still have been necessary. For Mayer's whole system reduces to the following hypothetical proposition: If work and heat are equivalent, then it follows . . . etc. To draw a conclusion, another premise is needed, namely: Work and heat *are* equivalent. Joule supplied that premise.

But Joule, knowing nothing of Mayer, also had to rediscover and put together at least as much of the major premise as would account for his immediate observations. Being a practical man, he didn't go much further. Others supplied the rest, and little by little the whole of Mayer's system was put together, and, being done by practised experts, was better rebuilt than it had been originally built.

A similar thing would have happened if the work of Copernicus had remained unknown. Galileo, when he turned his telescope on the skies in 1609, would have discovered facts that were at variance with the Ptolemaic system, such as the phases of Venus. He would himself have had to reconstruct at least so much of the Copernican system as would account for his own observations. As telescopes improved and observations accumulated, others would have added the rest, until at last the whole Copernican system would have been reconstructed, and better constructed than the original system.

Joule is famous chiefly for his accurate measurements of the mechanical equivalent of heat. That was of course a matter of great practical importance, especially for the power engineer. But from a theoretical standpoint his most important service was the establishment of the mutual convertibility of work and heat, which is the basis of the whole energy theory. His experiments with the magneto-electric machine, on the friction of fluids, on the compression and expansion of gases, and the free expansion experiment, showed that in every case the heat appearing or disappearing was *covariant* only with the work done, and not with any of the circumstances with which the calorists had connected it. That was the death blow of the caloric theory. What had been largely hypothesis with Mayer, Joule elevated to a fact. That was an invaluable service, and the sole honor of it belongs to Joule.

The evidence was completed, however, by Hirn in 1858, who showed that the heat of percussion is likewise proportional to the work expended, and in the same ratio that Joule had found for friction. [6]

To attempt to decide whether the work of Mayer or of Joule was the more important, is the same as to attempt to decide in general whether theoretical or practical work is the more valuable. The decision would depend upon the temperament of the judge. There are those who think that it is more important to have an automobile in which to fly about aimlessly, than to have correct notions about the universe. There are others who think the reverse. Both theoretical and practical work are necessary to the progress either of science or of civilization. That Mayer's work was necessary is proved by the fact that it all had to be done, and was done, over again. The ideas which he struggled in vain to put before the world, are now common school-boy knowledge, so familiar that most of us cannot remember a time when we did not know them. Joule helped us to the automobile, Mayer, or rather those who reconstructed him, helped us to a truer view of the universe. We should thank them both.

[6] *Recherches expérimentales sur la valeur de l'équivalent mécanique* (Paris).

CHAPTER XIV

PRECURSORS OF THE ENERGY THEORY

REVOLUTIONS in science as in other things do not consist in the sudden overthrow of a system in all its strength and glory. They occur only when the existing system is already tottering and only a little push is required to complete the overturn. The revolution is merely the final event in a long process of decay. This is necessarily so, for the revolutionists are always at first a minority, and weak.

Thus the Copernican revolution came when the Ptolemaic system was already overloaded to the point of collapse. The conservation of energy came when the caloric theory was so far gone that only the sharp thrust of Joule was needed to finish it. The more recent overturn in our physical conceptions came as the result of a long list of failures of the classical dynamics to account, for example, for the motion of Mercury's perihelion (Leverrier, 1848), for the lack of effect of motion through the ether (Michelson and Morley, 1881), and for all of the phenomena of radiation and of electricity that we now attribute to subatomic activities.

The conservation of energy when it came was long overdue. The experiments of Rumford and Davy in 1798 and '99 should have disposed of the caloric theory, but they didn't. It is not surprising therefore that the revolution broke out in several places at once. Naturally too, an event so long delayed had many precursors, many who helped prepare the way—many revolts that almost succeeded. As early as 1690, Huygens, as we have seen, declared that all parts of physics were branches of mechanics. The pronouncement was easy so long as little was known about these parts. But in the following century, the branches grew further and further apart and seemed to have less and less connection with one another. This was largely due to the habit of attributing each group of phenomena to a separate imponderable. There were no less than seven of these—phlogiston, caloric, light corpuscles or the ether (sometimes more than one ether), two electric fluids, and two magnetic fluids—to say nothing of the vital fluid, the bodily humors, the chemical spirits and essences, and other mysterious substances not connected with physics. It was the golden age of materialism.

But the nineteenth century, which has been characterized by Heyl as the century of correlation, [1] ushered in a reverse trend. Lavoisier had demolished phlogiston. Young and Fresnel disposed of the light corpuscles. Faraday deliberately set out to find connections between the most diverse phenomena. He found connections between chemical, electrical, magnetic, mechanical, and optical phenomena, culminating in his famous magnetization of a ray of light in 1845. He sought long but vainly for an effect of electricity on light. It was found by Stark in 1913.

In 1842, W. R. Grove expressed the opinion that all of these "affectations of matter" are forms of force, since each can produce mechanical effects, and can be produced by mechanical actions. Each is transformable directly or indirectly into any other; and always when one appears, another or others disappear. There are therefore as many dynamical equivalents as there are pairs of forces, and experiments, he said, should be made to determine them. He also suggested that all these "affectations" might be forms of motion. These ideas he expanded into a series of lectures, "On the Correlation of Physical Forces," which he delivered in 1843 and later published. [2] Grove thus approaches closely to the conservation idea, but never quite reaches the heart of the matter, which is the convertibility of work and heat.

On the other hand, in that same year, 1843, Colding, a Danish engineer, touched the heart of the matter in expressing the conviction that work is convertible into heat, and that in general

[1] Paul R. Heyl, *Fundamental Concepts of Physics* (Baltimore, Williams and Wilkins Co., 1926), Chap. II.

[2] W. R. Grove, *On the Correlation of Physical Forces* (1st American from the 4th English ed., New York, D. Appleton and Company, 1865).

mechanical force cannot be destroyed. [3] His reasons were so mystical, however, that Oersted suggested he make experiments. This Colding did, and from some two hundred of them on the friction of iron plates obtained for the mechanical equivalent 350 kilogram-meters (640 foot-pounds). But Joule meanwhile made more accurate measurements.

As early as 1839, Séguin, a French engineer, in a curious work on railways, [4] expressed the opinion that heat and motion are different manifestations of the same thing, and mutually convertible, and that steam after doing work in an engine delivers less heat to the condenser than it drew from the boiler. These ideas, he said, he obtained from his uncle Montgolfier, to whom they were suggested by a study of the hydraulic ram. Séguin also connected the temperature changes that occur when a gas is expanded or compressed with the work done. He gathered a good deal of data, but made no use of it until 1847, and then only at the suggestion of Joule.

Still earlier, in 1837, Mohr expressed the opinion that there were only two sorts of things in the world, matter and force, that the latter appeared in various forms, each convertible into any of the others, and that all were fundamentally forms of mechanical force or motion. He did not obtain any quantitative relations however. His paper has the distinction of having been refused by Poggendorff and was published in an obscure journal. [5]

But the earliest, most complete, and most remarkable of all the anticipators of the energy theory was none other than Sadi Carnot. We have mentioned that in 1824 he already had doubts about the "solidity" of the caloric theory. These doubts had become convictions by the time of his death in 1832, as we know from the notebook he left behind. Extracts from this were published by his brother in 1872, in connection with a second edition of the "Motive Power." Carnot there says that the experiments of Rumford and of others on friction and percussion, had convinced him that heat is transformed motion, perhaps a vibratory motion of the molecules. "If it is matter," he says, "it

[3] *Theses concerning Force,* Copenhagen Academy of Sciences.

[4] *De l'influence des chemins de fer* (Paris).

[5] *Ansichten über die Natur der Wärme, Baumgartner und v. Holger's Zeitschrift für Physik,* Vol. V.

must be admitted that matter is created by motion." Also he believes the converse conversion possible. He asks: "Is it absolutely certain that steam after having operated an engine and produced motive power can raise the temperature of the water of condensation as if it had been conducted directly into it?" In 1824 he answered yes, but now he answered no. In the end he concluded: "Heat is simply motive power, or motion which has changed form. It is a movement among the particles of bodies. Whenever there is destruction of motive power there is, at the same time, production of heat in quantity exactly proportional to the quantity of motive power destroyed. Reciprocally, whenever there is destruction of heat, there is the production of motive power. . . . There can never be either production or loss of motive power in nature. This power must be as unchangeable in quantity as matter."

Here we have the mutual convertibility of work and heat, the motion theory of heat, and the indestructibility of energy, combined as in the modern theory. But as though this were not enough, Carnot calculated the value of the mechanical equivalent. He does not give his method. He simply says: "According to some ideas that I have formed on the theory of heat, the production of a unit of motive power necessitates the destruction of 2.70 units of heat." This is really the heat equivalent of work. Reversing it, and converting Carnot's units into modern ones, we find that his value of the work equivalent of heat was 370 kilogram-meters, falling thus midway between the values obtained by Mayer and by Holtzmann. It was probably obtained by the same method and from the same data.

But Carnot was no mere speculator. He drew up a list of experiments by which his theory was to be tested. The list is no less remarkable than his theoretical anticipations, for it includes every essential experiment by which the conservation theory was actually established. Here are some of them, and the names of those who later carried them out.

"To repeat Rumford's experiments . . . but to measure the motive power consumed at the same time as the heat produced," substantially Joule's method, 1845–1849.

"To strike a piece of lead . . . to measure the motive power consumed and the heat produced, . . ." Hirn, 1858.

"To strongly agitate water . . . in a double-acting pump having a piston pierced with a small opening," Joule, 1843.

"To admit air into a vacuum. . . ." Gay-Lussac, 1807; Joule, 1845.

"To measure the changes which take place in the temperature of the gas during its changes of volume . . . comparing these with the quantities of motive power produced or consumed," Joule, 1845.

"Expel the air from a large reservoir in which it is compressed, and check its velocity in a large pipe in which solid bodies have been placed; measure the temperature when it has become uniform. See if it is the same as in the reservoir." This is Thomson and Joule's famous porous plug experiment, 1852, which we will describe later (Chapter XVI).

Had Carnot been granted a longer life, there is no doubt that the energy theory would have appeared much earlier than it did, and there would have been but one author, hence no unpleasant squabble over priority.

It is a significant fact that Carnot and all three of the founders of the energy theory were extremely young men, and that none was a professional physicist at the time of his discovery. Mayer never was one; Carnot was an engineer; Joule never held an official position, but remained all his life a private experimenter; Helmholtz did not obtain a chair in physics until twenty-four years after his discovery. These young men, unfettered by traditions, struck boldly out in a direction which the older professionals feared to take, and for long could not be persuaded to follow.

CHAPTER XV

ROOT OF THE CONSERVATION
OF ENERGY

WHILE MAYER'S attempted a priori deduction of the conservation of energy greatly antagonized the physicists of his time, there were others, especially in philosophical Germany, who saw in this very deduction the most valuable part of his work, and who, as Helmholtz put it, applauded him as a hero of pure thought. Indeed, the conservation of energy, once it is found, appears almost self-evident. It seems incredible that so much labor and battling should have been necessary to establish it. How indeed can we have a rational and unified conception of the universe without it? I myself remember as a boy, wondering how there could have been any physics at all before the conservation of energy. I thought it must have been all wrong—like alchemy. It was with great surprise that I learned later how recently the law had been established, and that a great deal of sound physics existed prior to it.

The explanation of the seeming self-evidence of the conservation law is that it has both a formal and an empirical side. The demand that *something* be conserved is an indispensable condition of scientific thought, and is, as we pointed out in Chapter VII, as old as such thought itself. The further demand that something connected with motion be conserved is also very old. But these formal demands alone can give no clue as to *what* is conserved. They only create an intellectual vacuum, that experience must fill. Who, for instance, could predict from pure thought alone, that the particular combination of feet and pounds that we call work, and the capacity for which we call energy, was the long sought for entity that is preserved? Once found, it is easy enough to say "of course."

Mayer was strongly convinced that the conservation of energy was more than an empirical discovery, that it had a formal side that gave it a higher than empirical authority—and in this he was right. He made only the mistake of supposing, because the content when found fitted so perfectly the form awaiting it, that the content could be deduced from the form. The general principles from which he endeavored to deduce it—from nothing comes nothing and the cause is equal to the effect—are *not* metaphysical principles. The fact that they were ancient and expressed in Latin does not make them metaphysical. They are perfectly valid expressions of the demands of rational thought and were felt and formulated by the ancients when they attempted to apply rational thought to the physical world. But nothing factual can be deduced from them. The one does not show *what* comes from *what,* and the other does not show in what sense the cause, when found, is equal to the effect. These must be discovered by experience. Mayer did not deduce his energy law from these general principles. He only showed that it was in conformity with them, that it satisfied these intellectual demands.

And Mayer was not the only one who felt that the conservation of energy had a higher than empirical authority. All three of the founders and many others had this feeling, and expressed it—with some curious results. The Danish engineer Colding argued that since forces were immaterial and spiritual they could not be mortal and perishable. Hirn, another engineer, regarded the conservation law as self-evident, and said that the "axiom" nothing comes from or to nothing, was his guiding principle. [1] It may seem strange that practical engineers should have indulged in such mystical expressions, but practical men take their basic theories on authority, and are always the most naïve and dogmatic.

The most astonishing statements of this sort came from none other than the practical and experimental Joule. In a lecture delivered at Manchester in 1847, [2] a few months before the

---

[1] Hirn, *Théorie mécanique* (Paris, 1865) , p. 4.
[2] *On Matter, Living Force, and Heat, Scientific Papers* (London, 1884) , Vol. I, p. 265.

Oxford meeting at which he met Thomson, Joule, after telling his hearers that they would be surprised to learn that until very recently the universal opinion had been that living force (kinetic energy) could be absolutely and irrecoverably destroyed at any one's option, went on to say: "We might reason, á priori, that such absolute destruction of living force cannot possibly take place, because it is manifestly *absurd* to suppose that the powers with which God has endowed matter can be destroyed any more than they can be created by man's agency; but we are not left with this argument alone, decisive as it must be to every unprejudiced mind." He then went on to show the empirical evidence for the law, doubtless for the benefit of the *prejudiced* minds. Similar remarks were made by him in his 1845 paper and elsewhere.

Mayer was roundly denounced for his so-called metaphysical deduction; but no one seems to have upbraided Joule for deducing the indestructibility of energy from the attributes of God! The fact is, the English physicists have always been proud of their orthodoxy, the Germans of their philosophical training. The one looked to theology, the other to metaphysics, for the super-physical authority which both felt that the conservation law possessed.

The third founder, Helmholtz, frankly based his whole system on an extra-mechanical principle—the exclusion of perpetual motion. That this is an extra- and even a pre-mechanical principle, is shown by the fact that all the fundamental laws of mechanics are essentially expressions of it, have in large part been derived from it, and owe their high certainty to it. It was used by Stevinus in 1585 to establish the law of the inclined plane, from which the law of the lever and all the rest of mechanical statics can be derived, [3] and to establish the laws of hydrostatics. It was used by Galileo to establish the law of inertia. [4] It was used by Huygens, by Lagrange, and by many others. That the exclusion principle does not derive its high certainty primarily from empirical evidence, is shown by the fact that when first used with so much assurance by Stevinus, there had not yet been any spectacular number of failures of

[3] Mach, *Die Mechanik*. Mott-Smith, *Principles of Mechanics Simply Explained*, p. 39.
[4] *Principles of Mechanics Simply Explained*, p. 59.

perpetual motion machines. The golden age of these contrivances came in the following centuries, when some smattering of elementary mechanics began to percolate among the masses. There had been centuries and centuries of failures of the alchemists to produce gold; yet that fact never deterred any one. Transmutation was never held to be *in the nature of things* impossible, until the doctrine of the immutable elements was adopted by chemists toward the end of the eighteenth century. But this doctrine lasted only a little more than a century. We now find transmutation perfectly possible—though not profitable. Whence comes then the super-empirical certainty that perpetual motion is *in the nature of things* impossible, and never will be accomplished?

This question was subjected to a searching analysis and answered by Ernst Mach in a lecture delivered in Prague in 1871, which Mach himself printed in the following year in pamphlet form,[5] because a previous article of his had been refused by Poggendorff. Mach finds that the principle of excluded perpetual motion is a special form of the law of causality, and therefrom derives its super-empirical authority. It applies not only to mechanics but directly to all physical phenomena, whether or not they are reducible ultimately to branches of Newtonian mechanics—which Helmholtz thought necessary. It is not identical with, but closely related to, the conservation law. The principal of causality, Mach finds, is clearly perceived at a very low stage of culture, though it may not be, and usually is not, correctly applied. Let us develop this line of thought.

If the conservation idea is as old as thought itself, the habit of seeking causes is still older. Even the animal, when it hears an unwonted sound, elevates its head, pricks up its ears, looks in the direction whence the sound came, and seeing perhaps some movement, goes finally to investigate. It does not of course say to itself—I will go and find out the *cause* of the sound and the movements. It simply goes. The whole action is a matter of in-

[5] *Die Geschichte und die Wurzel des Satzes von der Erhaltung der Arbeit* (Prague, 1872). English translation by Philip Jourdain, *History and Root of the Principle of the Conservation of Energy* (Chicago, Open Court Pub. Co., 1911). A reëlaboration is given also in Mach's *Popular Scientific Lectures*, trans. by T. J. McCormack (Chicago, Open Court Pub. Co., 1894). The subject is again discussed by Mach in his *Wärmelehre* (Leipzig, 1896), p. 315 ff.; and also in the latter part of the later editions of his *Mechanik*.

stinct and habit. Even man for the most part acts first, and thinks afterwards—if ever. This instinct and habit is very important for the survival of the animal. What it expects to find as the result of its search, is something alive. It may be a danger to be avoided, or an opportunity to be grasped. In either case, it is important to find out. Or it may be a chance for fun, or a matter of no interest. Animals, including primitive man, thus become expert in recognizing the *signs* of life, whether they be sights, sounds, or smells. They are perpetually seeking causes long before any man understands the nature of what he is doing. And the one cause they always seek—is life.

The prime sign of life is movement. Wherever primitive man saw any unusual movement, he suspected life, and often found it. Movement is psychologically exciting even today. There is nothing like it to attract, or to distract, attention. Characteristic of life also is the habit of concealment. Man did not always find the life he sought; but he never doubted its existence. Often things that seemed dead turned out to be alive. What looked like a dried leaf, or a twig, or even a stone, unexpectedly turned out to be a live creature. Some animals lie dormant for a long time, and then spring to life. Primitive man was thus forced to suspect life everywhere, even if he did not always find it. Everything was either alive, or might become so at any moment. This was the stage of universal animism.

It required a large amount of remembered experience to convince man that some things are *permanently* dead. But though the world thus became divided into the living and the lifeless, there was still but one recognized *cause* of every form of "liveliness," namely—life. The living creature can move and act of itself. The lifeless thing must *be* moved always by some living agency. This is the stage of limited animism.

Since much life exists which never meets the eye, it does not greatly surprise primitive man that he does not see the hand that sways the trees, or the creature that voices the thunder. It is a short step from unseen to unseeable life—to the ghost theory. Man is further led to this by his dreams, by death, and by many other experiences. Also the idea is attractive. Often man has to escape from his enemies by concealment, and is hard put to find it. What an advantage it would be if he could make himself invisible! What power it would give him over his friends and enemies! Unseeable creatures are therefore very powerful, and much to be feared. Those who can influence them are also very powerful. Thus arises priestcraft, with all its frauds and pretensions.

The next step is magic. Strange as it may seem, this is a step in advance and toward science. Magic admits the existence of *material causes*. She admits that one *thing* can affect another thing, without the intervention of any vital agency. The first cause is still animistic, but between it and the final effect may intervene a whole chain of material causes. Thus a man may push a stone over the edge of a mountain, and as it goes bounding down the slope, it starts other stones rolling, and these others, and so on. The existence of material causes was a very important discovery. It arose from a wider and closer observation of nature. But the observations were not yet sufficient to show in detail just what causes what and how. In the absence of this knowledge, the presumption was that anything might cause anything. Consequently the magician tried everything. What he tried and what he expected, seem to us of course very absurd. But his actions were so strange, not because his reasoning powers were in any way inferior to ours, but because he had not the material knowledge from which correct conclusions could be drawn. Such knowledge we have inherited from a vast recorded past experience. This knowledge is slow in accumulating, because primitive man is not interested in the connections between things *per se,* but only in bringing about certain desirable results, which he hopes to accomplish easily. It is only after innumerable failures that he undertakes the hard and not immediately profitable labor of finding out the real connections. Mach puts the matter thus: [6]

He who has no experience will, because of the complication of the phenomena surrounding him, easily suppose a connection between things which have no perceptible influence on one another. Thus, for example, an alchemist or a wizard may easily think that, if he cooks quicksilver with a Jew's beard and a Turk's nose at midnight at a place where roads cross, while nobody coughs within the radius of a mile, he will get gold from it. The man of science of to-day knows from experience that such circumstances do not alter the chemical nature of things.

[6] *History . . . of the Conservation of Energy,* trans. Jourdain, p. 64.

Extremely fantastic to us also is the idea of magic that anything can change into anything else or into nothing. But to superficial observation this is a most conspicuous and common occurrence. Things are continually changing—in size, form, color, and other properties. Thus ice changes to water, water to steam clouds, and the latter dissolve and disappear altogether. To the primitive man, these are not the same substance undergoing changes of state, but actual metamorphoses of one substance into another. This is evidenced by the different names that we still give to the different states of water. We do not do this with alcohol or with other substances. Similarly the Eskimo has different names for snow, according to whether it is in the air, spread out on the ground, drifting, or piled in drifts. And these names have no more etymological connection than our ice, water, and steam. [7]

Science, as we know, explains all change by the rearrangement of unchanging parts, and the direct qualitative changes of magic are ruled out. Only changes of position or of quantity are allowed, and everything must take place by a process, and according to a rule.

But the age of science has only just begun, and that of magic has not yet ended. The vast majority even of cultured people still have the habits of mind that pertain to magic rather than to science. They are ready to accept *any* cause in explanation of *any* effect, whether any connection, or any process by which the cause could bring about the effect, can be discovered or not. Many still believe that the moon affects the weather, that a certain kind of weather is favorable to earthquakes, that knocking on wood will influence the course of future events, and believe in a million other causal connections that do not connect. The prevailing conception of the universe still is that it began with a miracle and will end with a miracle, that only in the intervening stretch can science apply, and even this is not safe from supernatural interventions.

But magic is the stepping-stone to science, for she admits a chain of natural causes between miracles. The scientist is merely one who sets about examining this middle portion more closely. Gradually he lengthens and broadens it, and so enlarges the

[7] Franz Boas, *The Mind of Primitive Man* (New York, The Macmillan Company, 1913), p. 145.

domain that can do without miracles—the domain of science. In ancient times, a separate deity or angel steered the course of every star and planet. Aristotle, with the help of mathematicians, devised a system by which all the complicated motions of the heavenly bodies could be derived from the uniform rotation of a single outer sphere. Thus only a single deity was needed to turn eternally the crank of the universe—a great saving in celestial labor. (It is not known what became of the others who were thrown out of work.) Descartes did still better. It was only necessary to give his universe an initial push, a certain quantity of motion, which thereafter automatically conserved itself. Newton believed that God originally fixed the sizes and masses of the planets, the inclinations of their axes and orbits, started each with its proper velocity and spin, and left the rest to the law of gravitation. Laplace, with his nebular hypothesis and theory of probability, did away with the need of this original fixing, and when Napoleon asked him where in his universe there was a place for God, Laplace answered that he had done without that hypothesis.

Science, however, does not abolish the magician. She merely devotes herself to that realm which can do without him, where material causes hold full sway. Here she finds, according to Mach, that groups of phenomena depend upon other groups. They vary together, or vary inversely. When A appears, B appears; or when A *disappears,* B appears. We then say that A is the cause of B. But in this realm we demand, not only that every effect shall have a cause, but further that the cause shall be adequate to the effect (the law of sufficient reason), or is in some sense *equal* to the effect. This simply means that cause and effect must be connected by a fixed *quantitative* relation. When A increases, B increases; or when A *decreases,* B increases. In either case, the *value* of B is *determined* by that of A. Mathematically speaking, we say that B is a *function* of A, which simply means that the relation by which A and B are connected, can be expressed by a mathematical equation. This equation gives both the manner in which B varies with A, the *form* of the function, and the values of B corresponding to every value of A. There are as many such possible relations as there are equations that we can write—an indefinite number. The particular one that obtains in a given case must of course be found from experience (by measurements), or deduced from relations already established. Because of this functional relation, the variations of B follow exactly those of A. If A is constant, so is B; if A varies, so does B. If A after a series of changes returns to its initial value, B likewise returns to its initial value. This last statement, according to Fechner, contains the kernel of the whole law of causality.

We have for simplicity supposed above that B is completely determined by A. But in general, B will be dependent upon several factors (will be a function of several variables). Thus the volume of a gas depends both upon its temperature and its pressure, and on other minor factors. Conversely, the variations of one thing may affect several other things. Hence in general we must say, one group of phenomena A, B, C . . . affects another group A', B', C'. . . . It may be that everything affects everything else—somewhat. But the vast majority of effects are so extremely slight that they may be disregarded, and attention paid only to the major ones. Besides, it is often possible to screen off, to balance, or to compensate factors that are not wanted— that is, to isolate a phenomenon. It is also possible to vary one factor at a time, keeping the others constant, and study its separate effect. Thus we can keep the temperature of a gas constant, and study the variation of its volume with the pressure; or we can keep the pressure constant, and study the variation of volume with the temperature.

One may ask—how do we know that nature must or will conform to these intellectual demands? I answer—we don't. We study her only in so far as she does. If the scientist encounters anything that does not conform, he turns aside. It does not belong in his domain. He seeks elsewhere for things that do conform—and finds plenty of them. The beautiful conformity then that we observe between thought and things in science is due to the fact that the things are selected because of this conformity. That we find so many of them, that the domain of science is so large, suggests that there is a real and inherent conformity between thought and things. But if nature has produced the mind, it is not surprising that the processes of the mind should parallel those of nature, that the variations of the one should be a function of those of the other. If, conversely, the mind has created the physical world, as some think, it has of course created it in its own image. Finally, if the mind has been specially created by a beneficent supernatural Being, for what purpose should it be created but to help the creature orientate itself in its environment? By all three theories a correspondence is to be expected. This may not be very profound philosophy, but it is a very natural thought.

Finally, Mach concludes, the functional relation means that the variations of A and B must be like in kind. If A varies periodically (goes through a cycle), so does B. If A changes continuously in one direction (grows continuously larger or smaller), so does B. *A periodic variation in A cannot produce a continuous change in B.*

This last statement, says Mach, is equivalent to the exclusion of perpetual motion. For a perpetual motion is nothing other than an attempt to get from periodic variations in one system of bodies, continuous changes in another. Take, for example, the simplest device, the overbalanced wheel of Figure 17. After each revolution, the wheel is in exactly the same condition that it was at the beginning. It undergoes only periodic variations. Yet it is claimed that this wheel will do mechanical work, that is, produce continuous changes in other bodies. In that case we could set it to turning a grindstone, and by pressing objects against the latter, could gradually wear down both the stone and objects until they were reduced to dust. We would then have utterly consumed the stone and the objects, but the wheel—disregarding its own wear—is as fresh as ever—ready to eat up another grindstone, and as many more of them as we might offer. The stone has suffered a permanent and irreparable change, but the wheel has suffered none. A continuous effect, however, requires a continuous cause. There is no such cause here. Hence according to our principle, the effect is totally uncaused. It is nothing less than a miracle. It does not belong to the domain of science.

Suppose, on the other hand, we turn our grindstone by an electric motor operated by a battery. Then the permanent change in the stone is matched by a permanent change in the battery. As the stone is consumed, zinc is consumed. We can equate these two like quantities. But we cannot equate the permanent change in the stone to the periodic variations of the wheel, any more than we can equate cabbages to kings.

It is a significant fact that no perpetual motionist ever states the horse-power of his machine. He doesn't start out as other engine builders do, to build a one- or a two-horse-power perpetual motion machine. The very idea sounds ridiculous. He hopes to get *some* power, but how much he will find out when and if his machine runs. Nor does he ever give a formula by which the power could be calculated from the dimensions of the machine or from any other quantities involved. He can't, for all continuous changes are zero, and that is the only nonperiodic quantity involved to which the horse-power could be equated—and he will not admit that zero is the right answer.

The exclusion of perpetual motion is thus a consequence of the functional relation of cause and effect. But it is just this functional relation that makes *exact* science possible. Without it, no quantitative laws could be established, no quantitative predictions could be made, any more than the horse-power of a perpetual motion machine can be calculated or predicted. The domain of exact science would shrink to nothing, and nowhere would we be safe from miracles and magic.

But this causal law is not handed down from heaven. It is not an a priori principle, an innate idea, or an intuitional insight. It is an empirical discovery. But it is derived from and is exemplified by *all* experience with the physical world. It has therefore a higher authority and assurance than any special or limited experience, or even than any considerable amount of the latter, such as several centuries of failure of perpetual motion. Because the exclusion principle is part and parcel of the causal law, it shares the latter's seemingly super-empirical authority.

The law of the conservation of energy is a fulfilment of the demand that the cause be equal to the effect in the case of continuous or permanent changes, just as Mayer insisted. But before this demand could be fulfilled, a way had to be found by which the cause and the effect could be measured and thus equated. This was accomplished when it was found that heat was transformed work, and that all the other so-called forces of nature could be measured in terms of work—of pounds multiplied by feet. Work is thus the common factor that enables all these forces to be compared, that unites all branches of physics.

Consider, for example, our grindstone turned by an electric motor. As the stone is worn down, zinc is consumed. Undoubtedly the two are causally connected. Both are continuous changes and can be equated. But we shall not find an equation that will apply in all cases, because the quantity of material worn from the stone depends upon a variety of factors that pertain only to that particular stone. Similarly with the amount of zinc eaten up in the battery. But, independent of all these special circumstances, the *heat* developed by the oxidation of the zinc, after deducting what may be otherwise expended, is always equal to the *work* done in wearing down the stone. The equivalence of work and heat thus enables us to set up a universal equation, the energy equation.

In reality the mechanical equivalent of heat is not a relation between two different things. It is a relation between two sets of units by which we measure the same thing. We invented the calorie before we knew that heat was transformed work and therefore properly measured in foot-pounds. Similarly we invented electrical units before we understood the nature of electrical energy. There is also a relation between them and mechanical units. Everything could and should be measured in energy units, but we retain our calories and our watts because we are used to them.

The conservation of energy is connected with perpetual motion because the latter pretends to create work. If all resistances were removed, a wheel once started would turn forever. This would be a perpetual motion, but not in the usual sense, because it does no work. It produces no effect on the rest of the universe. It is completely isolated. Such complete isolation is of course never actually possible. The wheel must overcome at least the friction of its own axle. This continuous work requires a continuous cause, and the only continuous change that can occur in the wheel is a decrease in its speed. This can go on only until the speed is reduced to zero. Only a limited amount of work can thus be obtained from an isolated system. If there is to be a continued outflow, there must also be a continuous inflow of energy.

Another illustration is given by Mach: A perfectly elastic tuning fork vibrating in a perfect vacuum, would continue its oscillations forever with undiminished amplitude. But no sound would be heard. In order for a sound to be heard, the vibrations must be received by a medium and transmitted to the eardrums.

In short, the vibrations must be damped, their amplitude must diminish, and the sound continues to be heard only so long as the vibrations continue to diminish. For the sound to be indefinitely prolonged, the vibrations must be kept up by renewed impulses from without. Mayer compared the sun to an ever-sounding bell, signifying that its vibrations must be continually renewed from some source of energy.

The conservation of energy not only denies that continuous effects can flow from periodic causes—perpetual motion—but asserts further that whenever mechanical work is created or consumed, there is a transformation of energy from one form to another, and that always the amount of the one is *equal* to the amount of the other when both are measured in the *same* units—say foot-pounds—so that the sum total of all the energy in the universe is a constant. In short, energy—not heat or work—is indestructible and uncreatable. The conservation of energy thus takes its place beside the conservation of matter. Recently we have combined the two. Energy and matter appear to be mutually convertible in a fixed ratio, so that we now say that the total matter *and* energy of the universe is a constant.

CHAPTER XVI

THE PRINCIPLES OF CARNOT AND OF JOULE UNITED—CLAUSIUS, 1850

THOMSON, we have seen, could not bring himself fully to accept the views of Joule because they conflicted with the theory of Carnot. According to Joule heat is transformed into work, and thereby ceases to exist as heat. According to Carnot work is produced solely by the fall of heat without any alteration in its quantity. In 1849, Thomson wrote an account of Carnot's theory in which he still adhered to this view. [1] But in the following year, Clausius [2] showed that the two principles could be brought into conformity by means of a surprisingly small modification of the theory of Carnot. The heat drawn from the source, said Clausius, may be regarded as composed of two parts, one of which is transferred to the cooler, the other converted into work and destroyed. There is nothing in this to conflict with the essential part of Carnot's theory, which is, that the motive power developed in a perfect cycle depends only on the quantity of heat *drawn from the source*, and the temperatures of the source and cooler. It is by no means essential to this theory that the quantity of heat delivered to the cooler should be the same as that drawn from the source.

Joule's discovery, the convertibility of heat and work, Clausius called the first law of thermodynamics. Carnot's theory, he said, is founded on the proposition that heat cannot of itself pass from a colder to a hotter body. This he called the second law of thermodynamics. From these two laws he not only proved that the work obtainable in a perfect cycle from a given quan-

[1] *Transactions of the Royal Society of Edinburgh,* Vol. XIV.
[2] *Poggendorff Annalen,* Vol. 79, 1850.

tity of heat drawn from the source depends only on the temperatures, but also showed the exact manner of this dependence, which Carnot had left undetermined. In short he evaluated the Carnot function (page 40 above). This he did by "applying" the two laws to the Carnot cycle both for perfect gases and for saturated vapors.

The first law was applied to the gas cycle with the help of the following considerations, which though not new were given significance by the use that Clausius made of them. When a gas expands adiabatically, the temperature falls because some of the heat in the gas is converted into work. To prevent the temperature from falling, that is, to expand isothermally, precisely this same amount of heat must be restored to the gas. Hence all of the heat absorbed during an isothermal expansion is converted into work. It simply passes through the gas—in as heat and out as work—leaving the heat content of the gas unchanged. During the first operation, then, of the Carnot gas cycle, the isothermal expansion ab, Figure 18, all of the heat drawn from the source, which we shall call Q_1, is immediately converted into work; and this work is represented by the shaded area on the diagram. Hence we can at once determine its amount, either by measuring the area, or by calculation from the equation of the isothermal. Dividing it by the mechanical equivalent of heat, we obtain Q_1, the quantity of heat drawn from the source.

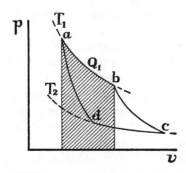

FIG. 18. WORK UNDER THE HIGHER ISOTHERMAL

The amounts of work done along the two adiabatics bc and da are represented by the shaded areas shown in Figure 19. The work under bc is done *by* the gas at the expense of some of its own heat, with a consequent drop in temperature. The work under *da* is done *on* the gas, adds to its heat content and raises its temperature. But the two adiabatics lie between the same pair of isothermals. The temperature is raised along *da* by the same amount that it was lowered along *bc,* and since the specific heat of the gas does not change with the volume (as shown by the free expansion experiments) nor with the temperature (as shown by the measurements of Regnault) the heat involved in each case is the same. So also is the work. The two shaded areas are equal. The adiabatics cancel out. They are mere temperature changers.

Finally, the work done on the gas during the isothermal compression *cd,* is shown by the shaded area in Figure 20. It is immediately converted into heat, which we shall call Q_2 and

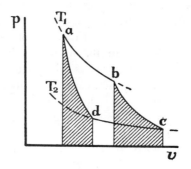

Fig. 19. Work under the adiabatics

rejected to the cooler. It is evident from the diagrams that Q_2 is less than Q_1. The heat rejected to the cooler is less than that drawn from the source. The difference has been converted into work, and this work is represented as always by the area enclosed by the cycle. Since no change in heat content takes place along either isothermal, and the opposite changes along the adiabatics cancel, the gas, when it again arrives at *a,* has the same heat content as at the start. Thus Carnot's requirement that the working substance undergo no permanent change is also satisfied. These are the results of the application of the first law.

To apply the second law, Clausius made use of an infinitesimal cycle. We shall follow his later demonstration which applies to a finite cycle. [3] The two isothermals *ab* and *cd* lie between the same pair of adiabatics. Having determined the equation of the adiabatic curve, Clausius was able to prove that the heat absorbed or rejected along any such pair of isothermals is proportional to the absolute temperature at which it is absorbed or rejected. This result is epitomized by the following proportion

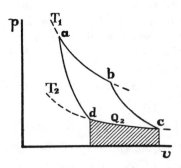

FIG. 20. WORK UNDER THE LOWER ISOTHERMAL

$Q_2/Q_1 = T_2/T_1$, where T_2 and T_1 are the absolute temperatures of the lower and upper isothermals respectively. Subtracting the number 1 from both sides of the equation, we get $\dfrac{Q_1 - Q_2}{Q_1} = \dfrac{T_1 - T_2}{T_1}$. Now $Q_1 - Q_2$ is the amount of heat we get turned into work, and this we see from the equation is proportional to Q_1, the quantity of heat drawn from the source, and to the difference in temperature between the source and the cooler, to the "fall of heat," as Carnot said. But it is also inversely proportional to the absolute temperature of the source, since T_1 occurs in the denominator. This is new. It is the long sought evaluation of Carnot's function—the law according to which the amount of work obtainable from a given quantity of heat and a given temperature fall, diminishes as the temperature of the source is raised. It turns out quite simply that the work obtainable is inversely proportional to the absolute temperature of the source.

We have in Chapter IV compared this diminution in the "force of heat" to the diminution in the force of gravity. Now

[3] *Die mechanische Wärmetheorie* (2nd ed., Braunschweig, 1876). Vol. I. p. 85.

gravity varies inversely as the *square* of the distance from the center of the earth—a much higher rate. But we are much further from the absolute zero of gravitational force—the center of the earth—than we are from the absolute zero of temperature; and the differences in levels we have to do with on the earth are very small in comparison with the distance to the center. Hence the variations in gravitational force are quite imperceptible. The differences in temperature we deal with are, on the other hand, large in comparison with the drop to the absolute zero, so that the variations in motive force with the temperature height are more conspicuous.

Since Q_1 is the heat drawn from the source, it is what we pay for on our fuel bill. $Q_1 - Q_2$ is the portion of it that we get turned into work. Hence $\dfrac{Q_1 - Q_2}{Q_1}$ is the ratio of what we get to what we pay for. It is the *efficiency* of the cycle. It is equal to $\dfrac{T_1 - T_2}{T_1}$, to the ratio of the temperature drop to the absolute temperature of the source. Obviously this fraction will always be less than 1, the efficiency always less than 100 per cent, unless $T_2 = 0$. We cannot get all of the heat drawn from the source turned into work unless the cooler is at the absolute zero. Since this is impossible, even a perfect engine will always have an efficiency less than 100 per cent, usually much less. The conversion of heat to work is thus at the very outset loaded with a theoretical handicap, whatever may be the further losses in the practical engine.

In 1851, Thomson, at last converted to the new theory, wrote a paper on the subject himself. [4] In this he proved in a different way all the results that Clausius had obtained, and furthermore showed that what Clausius had so far proved only for an infinitesimal cycle was true also for a finite cycle. He also drew some far-reaching conclusions. Clausius excelled in mathematics, Thomson in clear and simple statement. It has been said of Clausius that one never knew whether he was trying to say something or to conceal something. What he endeavored to conceal in fact was how he actually got his results. He made tremendous efforts, especially at first, to prove everything without

[4] "On the Dynamical Theory of Heat," *Transactions of the Royal Society of Edinburgh*, Vol. XX, Part II.

assumptions. Consequently his early methods were tediously roundabout and obscure. His later writings, after certain essential assumptions had become generally accepted, were much clearer. Thomson from the start was much more direct and concise, and knew how to express his results in words that could be understood. He became the interpreter and expounder of the theory. It is doubtful if any but the mathematicians would have understood it without him.

Thomson was much impressed by the fact that only a part, and usually a small part, of the total heat drawn from the source could be converted into work, "the remainder being irretrievably lost to man, and therefore 'wasted' although not annihilated." For this part having descended to the temperature of the surroundings can never again be reconverted to mechanical work. On the contrary, work is required to lift it. It has become totally "unavailable." Every transformation of heat to work, even under perfect conditions, is thus unavoidably accompanied by the transformations of another quantity of heat from an available to an unavailable form.

This brings us face to face with a remarkable peculiarity of heat energy, which distinguishes it from all other forms. When by friction or otherwise we convert work to heat, we eventually convert all of it. The efficiency of this operation is 100 per cent. Moreover it takes place of itself; it is a *natural* operation—like the flow of heat down-temperature. The conversion of heat to work, and the raising of heat to a higher temperature, on the other hand, are hard uphill jobs. They must be paid for, the one by letting other heat down-temperature, the other by converting mechanical energy into heat. If by the one we have gained mechanical energy, this *when used* is eventually all reconverted into heat. If by the other we have raised a quantity of heat up-temperature, it eventually all flows down again to the general level. These hard uphill jobs produce no permanent results. In the end everything goes downhill.

Moreover all other forms of energy can be converted one into the other at *par*. Thus, electrical energy can be converted into mechanical energy and *vice versa* with theoretically 100 per cent efficiency. Each of these other forms of energy can also be converted into heat with 100 per cent efficiency, but *not vice versa*. Once in the form of heat, energy can be converted to these other forms only partially and temporarily. We must therefore look upon heat as *low grade* energy, and the lower the temperature the lower the grade. The other forms are *high grade*. There is, hence, a universal trend to the *degradation* of energy.

In the following year Thomson developed this thought further and in a different way.[5] High temperature heat, he said, is concentrated heat; low temperature heat is dispersed heat—for a quantity of heat that will raise a small body to a high temperature must be communicated to a large body or to many bodies in order to raise their temperatures only insensibly. The flow of heat down-temperature is hence at the same time a dispersion or a dissipation of heat. And since all forms of energy when used eventually degrade into heat, there is a universal trend in nature to the dissipation of energy. This is conspicuously shown by the stars, which are continually radiating enormous quantities of heat into space in all directions, and this heat is traveling away at the rate of 180,000 miles per second. The final result can only be that all the energy of the universe will be ultimately reduced to the dead level of uniformly diffused heat. All temperature differences will be wiped out, and no energy transformations can take place. All life will have long since been extinguished. The universe itself will be dead. This is the famous *heat-death* visioned by Thomson.

It is indeed a gloomy picture. Yet to some it is a cause for rejoicing, for in it they see a proof of the existence of God. Others, horrified, endeavor in some way to circumvent this consequence of the second law of thermodynamics, in order to preserve the *status quo* forever. It is a matter of temperament. Incidentally, there have been those who in the first law of thermodynamics—the indestructibility of energy—have seen a proof of the immortality of the soul.

In previous chapters we have discussed perpetual motion devices that conflicted with the first law of thermodynamics. There is another sort that does not conflict with this law—because it does not attempt an actual creation of energy—but that conflicts with the second law. We call it a perpetual motion of the second sort. To get work from heat, it is not sufficient to have a quantity of heat on hand. We must have also a difference of temperature.

5 "On a Universal Tendency in Nature to the Dissipation of Mechanical Energy," *Proceedings of the Royal Society of Edinburgh,* 1852.

Two things are necessary, a quantity and an intensity factor. The same is true of all forms of energy. To obtain work from gravitational forces, it is not enough to have weights on hand. They must be *raised* weights. There must be a difference of levels between which the weights can descend. To obtain work from electricity, there must be a difference of electrical potentials, so that the electricity may flow from the higher to the lower potential, etc. A perpetual motion of the second sort is an attempt to get work from energy on hand, without the requisite difference in the intensity factor. It is an attempt to get work out of weights that are already lying on the ground, out of water that is already at sea level, out of heat that is already at the temperature of the surroundings, out of a permanent magnet, etc. A famous imaginary example of the last is the floating island of Laputa in *Gulliver's Travels,* which could rise, sink, stand still in the air, or transport itself in any direction—the motive power being a huge magnet in the center of the island. Another device is a wheel in which the heat developed by its own friction is to be converted into motive power to drive the wheel.

There is indeed enough energy lying around to do all the world's work, if we could circumvent the second law. For example, the amount of heat contained in a pint of water at ordinary temperature, could, if all converted into mechanical energy, raise that pint of water to a height of sixty miles. We do not know the exact heat content of any body, but the above is based on a conservative estimate. To get all of this energy, however, we would have to provide a cooler at the absolute zero, and to provide this would entail the expenditure of far more energy than we could get from the water. We have a similar situation with gravitational energy. We can get work from a descending weight only until it reaches the ground. But this is by no means all the gravitational energy it contains. We could get the whole only by allowing the weight to descend to the center of the earth. But first we must dig a hole. We must *lift* all the material that occupied the place of the hole to the surface. This of course would entail far more work than could be got from the descent of the weight.

The fact that the efficiency of a Carnot cycle depends only upon the temperatures between which it works, and not at all upon the nature of the working substance, suggested to William

Thomson, as early as 1848, the possibility of setting up an absolute scale of temperature that would be independent of the properties of any material substance.[6] The various gas thermometers, he pointed out, had been shown by the accurate experiments of Regnault to agree very closely among themselves, but after all they represented only "an arbitrary series of numbered points," based on the assumption that the expansion of a gas is very closely proportional to the rise in temperature. Thomson proposed to make the degrees of his scale such that the fall of a unit of heat in a Carnot cycle through one of these degrees, would always produce the same amount of work. That is, the ratio of work to heat, or the efficiency of the cycle, was to be the same for a fall of one degree in any part of the scale. He knew that, according to Carnot's function, this ratio diminished as the temperature rose, so that the degrees determined in this manner would grow steadily larger in comparison with those of a gas thermometer, as the temperature rose. This would be an inconvenience, but since at that time the form of Carnot's function was not known, it could not be avoided.

After Clausius had determined the form of the Carnot function in 1850, Thomson perceived that to make his degrees really equal, he must make the work by which they are measured diminish in proportion as the temperature rose. They should then agree very closely with those of the gas thermometer.[7]

How to compare the two scales experimentally became a difficult problem. The Carnot cycle, being ideal, cannot be carried out in practice. Even if it could, we would have to measure quantities of work and of heat, and for the latter would have to use a thermometer of the old style. It was obvious that instead of measuring whole degrees and comparing them, a method must be found by which the *differences* between the degrees on the two scales could be directly measured. Any error then made by using an old style thermometer would be only a small part, say a hundredth, of this *difference*, which itself might be only a hundredth part of the whole degree. The error in the latter would then be but one part in ten thousand.

---

[6] *On an Absolute Thermometric Scale.* . . . *Philosophical Magazine,* 1848, Vol. 33, p. 313.

[7] "On the Dynamical Theory of Heat," Part V, *Transactions of the Royal Society of Edinburgh,* 1854, Vol. XXI, p. 213.

It was well known that a perfect gas would be a perfect temperature measurer, because its expansion is in exact proportion to the temperature rise. This is simply a consequence of the assumption that there are no forces of attraction between the molecules. If in a real gas there are such attractions, they will prevent the gas on being heated from expanding as much as it otherwise would. The heat we communicate to it will not all go to increasing the kinetic energy of its molecules—to raising its temperature. Some will go to increasing their potential energy, to the work of separating the molecules against their attractions, to the internal work. And a gas is an imperfect temperature measurer in proportion to the heat absorbed for internal work. When a gas expands into a vacuum, no *external* work is done, and there is no drop in temperature from this cause. But there should be a drop in temperature because of the *internal* work done. Undoubtedly it is very slight, so that Joule's free expansion experiment did not show it. Thomson felt that this matter should be tested by a more accurate method, and after considerable thought devised his famous porous plug experiment, which he and Joule carried out together.

This experiment consists in forcing a gas through a pipe in which there is an obstruction, the porous plug, by which its pressure is reduced, and in measuring the temperature on both

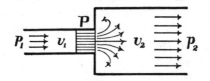

FIG. 21. THE POROUS PLUG EXPERIMENT

sides of the obstruction. It is the same experiment that Carnot proposed in 1824; but this was not known to Thomson or Joule. The pipe is heavily insulated so that no heat enters or is lost during the experiment. The arrangement is shown diagrammatically in Figure 21. The gas approaching the plug P in the smaller pipe under the pressure p_1, has its pressure reduced after passing the plug to p_2. This means that its volume increases, say from v_1 to v_2. Hence we have essentially an expansion experiment, and the pipe is appropriately enlarged beyond the plug. In issuing from the larger pipe, the gas must do work against the pressure p_2. This work is proportional to $p_2 v_2$. In forcing the gas through the plug, the work $p_1 v_1$ must be done on it. But according to Boyle's law, $p_1 v_1$ is equal to $p_2 v_2$ (and if the gas does not obey Boyle's law exactly, a small correction is applied), so that the work done on the gas is equal to the work done by the gas. In short, the gas does no external work of itself, that is, none at the expense of its own heat. And since no heat enters or leaves it, its heat content does not change. The experiment is therefore equivalent to a free expansion experiment. If cooling occurs, it must be due solely to the internal work, and a measure of it.

The experimenters found such a cooling, amounting in the case of air at ordinary temperature to a third of a degree Fahrenheit per atmosphere drop in pressure. The effect increased as the initial temperature was lowered. It was greater also for the less perfect gases like carbon dioxide. The first experiments were made in 1852, merely for the purpose of testing more accurately Joule's free expansion experiments. Afterwards they were applied to the correction of the gas scales in accordance with the thermodynamic scale, and were continued for several years.

The corrections turned out to be very small, amounting only to hundredths of a degree over the whole range for which gas thermometers are used (− 400 to + 800 degrees), and only to thousandths of a degree between the freezing and boiling points of water. Aside from the practical value of the thermodynamic scale, the cooling of a gas on free expansion was an important discovery, as we shall see.

CHAPTER XVII

ENTROPY

IN 1854, Clausius gave a new interpretation to the second law of thermodynamics, in the course of which he introduced a new concept, which has played a dominant and somewhat mysterious rôle in the theory of energy ever since.[1] In Carnot's theory, heat passes from the source to the cooler without any alteration in its quantity. In Clausius's theory, a part of this heat is destroyed, and a lesser quantity is delivered to the cooler. Nevertheless, Clausius thought, something must remain unchanged in the perfect cycle, and distinguish it from all others. What this something was, he believed was shown at once by the fundamental equation of the cycle, $Q_2/Q_1 = T_2/T_1$, which we gave on page 153. Transposing terms, this may be written $Q_1/T_1 = Q_2/T_2$. Expressing it in words, the heat drawn into the cycle divided by the absolute temperature at which it is drawn in is equal to the heat rejected divided by the absolute temperature at which it is rejected. It is this ratio of the heat transferred to the temperature at which it is transferred, that remains unchanged at the end of a perfect cycle.

This ratio of flowing heat to temperature Clausius regarded as a distinct thermal quantity, as significant as heat or temperature —just as in mechanics we have distance and time, and their ratio, velocity, is also a distinct and significant quantity. He therefore gave his ratio a name. At first he called it the transformation-content (*Verwandlungsinhalt*), because it measured the transformability of the heat. Later he called it *entropy*, which comes from two Greek stems meaning—turning into. This

---

[1] *Ueber eine veränderte Form des zweiten Hauptsatzes.* . . . . *Poggendorff Annalen*, 1854. The above exposition, however, is taken from his later and clearer writings.

quantity increases when heat flows into a body, decreases when heat flows out, and remains constant during a reversible process involving no flow of heat. The entropy of the working substance therefore increases continually along *ab*, Figure 18, page 151, as heat flows into the gas, until at *b* it reaches the value Q_1/T_1. It remains constant along the adiabatics because there is no flow of heat. Along the lower isothermal, the entropy decreases continually as heat flows out, until at *d* the quantity Q_2/T_2 has been rejected. This is the same amount that was taken in along the upper isothermal. Hence the entropy of the working substance is unchanged at the end of the cycle. The entropy of the heater, on the other hand, has been decreased by Q_1/T_1, and that of the cooler increased by the same amount. There is hence no change in the total entropy of the system. In the Carnot theory, a certain quantity of heat is transferred, unaltered in amount, from the heater to the cooler. In the Clausius theory, a certain quantity of entropy is transferred, unaltered in amount, from the heater to the cooler. We can therefore reinstate the water-wheel analogy, if for heat we substitute entropy. For this reason Zeuner called entropy—*heat-weight*.

No change in the total entropy of the system is thus the sign of a perfect cycle. Consider what happens if we put the heater and the cooler in direct contact. Heat then flows straight from one to the other, and no motive power is produced. Suppose that a quantity Q thus passes. The entropy of the heater is diminished by Q/T_1; that of the cooler is increased by Q/T_2. Now the denominator of the second fraction is smaller than that of the first one, while the numerators are the same in both. Hence the second fraction is larger than the first one. The entropy of the cooler has increased more than the entropy of the heater has decreased. Hence the entropy of the system has increased. It is obvious that such an increase will always occur whenever heat flows from a higher to a lower temperature. There is no change in entropy only when heat flows between bodies that are at the same temperature. But this, we saw, was precisely Carnot's condition for a perfect cycle.

Since heat never flows without some difference in temperature, it follows that all natural processes, all actual cycles, are accompanied by an increase in entropy. Everywhere then, entropy is on the increase. Energy is losing its transformability. Clausius therefore summed up the import of the two laws of thermodynamics as follows:

1. The energy of the universe is a constant.
2. The entropy of the universe tends to a maximum.

So far, we have considered the entropy change only along an isothermal. This is easy to calculate because the temperature is constant. We have only to divide the total heat absorbed or rejected by the constant temperature at which the operation takes place. But when we are concerned with a flow of heat during which the temperature changes, the matter is not so simple. Suppose the operation is represented by some curve on a *pv* diagram. Clausius showed that we can substitute for the actual path a flight of stairs composed of alternate bits of isothermals and adiabatics, the former composing the treads, the latter the risers. The total change in entropy is then the sum of all the changes that take place along the isothermals, there being none along the adiabatics. Now by making the steps smaller and smaller, we can approximate to any path. And if we know the equation of the curve, we can by integration find the change in entropy exactly. The process is the same as that for finding the work under any curve, which we described in Chapter III. Clausius showed that the change in entropy between any two given points is independent of the path taken, and depends only on the positions of the points, that is, on the initial and final states of the substance, just as we showed was the case with energy changes. It follows that the net change in entropy of a substance carried around *any closed* cycle, is zero.

Every heat-engine works on a closed cycle. There is no permanent change in the entropy of the working substance, any more than there is in its heat content. The increase in entropy consequent upon the cycle not being perfect, occurs in the rest of the system—in those processes by which we make up the losses of the engine.

We do not know the total or absolute value of the entropy of a body, any more than we do its heat content. We can only measure *changes* in these quantities. Consequently we have to measure entropy as we do heat up from an arbitrary zero. We choose the freezing point of water and atmosphere pressure as the state of a body in which we say its entropy and its heat content are zero. Thus at 32 degrees Fahrenheit, the entropy of a pound of water is zero. As heat is added, its entropy increases and the temperature rises. Since the latter rises continuously, to find the change in entropy, we must divide the process into a series of small steps. Let us first heat the water two degrees. This will require two therms. The average temperature during the operation will be 33 degrees, or on the absolute scale, 493 degrees. The increase in entropy is therefore 2/493 or .00406, and this is the entropy of a pound of water at 34 degrees Fahrenheit. Now let us heat the water two more degrees. This will require two more therms at an average temperature of 495 degrees absolute. The further increase in entropy is 2/495 or .00404, and this added to the previous increase, gives the entropy at 36 degrees as .00810. And so we may go on up to any desired temperature. But the process is tedious, and besides the variation in the specific heat of water must be taken into account. Hence tables have been prepared, the steam tables, which give the entropies of water, steam, and other substances, in all the conditions which are likely to occur in engineering practice. From such a table we find that the entropy of water at 212 degrees is .3118. This is called the entropy of the liquid. If now we boil the water, we must add 970 therms—the latent heat. Since the temperature remains constant during the boiling and equal to 672 degrees absolute, the increase in entropy is 970/672 = 1.445. This is called the entropy of vaporization. Adding this to the amount previously found gives the *total* entropy of the steam as 1.757. These three quantities correspond, the reader will note, to the heat of the liquid, the heat of vaporization, and the total heat of steam.

Since the entropy is constant along an adiabatic, each such line represents a particular value of entropy, just as each isothermal represents a particular value of temperature. Let us then designate the entropy of the first adiabatic *ad,* Figure 18, by N_1, that of the second by N_2. Then the increase in entropy as we pass from the first to the second adiabatic, by any path whatsoever, is $N_2 - N_1$. Now we can make a plot of temperature and entropy just as well as we can of pressure and volume. If we plot the temperatures vertically and the entropies horizontally, every isothermal becomes a horizontal straight line, every adiabatic a vertical straight line, and the Carnot cycle becomes the rectangle *abcd* of Figure 22. The increase in entropy along the upper isothermal, we have seen, is Q_1/T_1, and this according to our present convention is equal to $N_2 - N_1$. Transposing, we get $Q_1 = T_1 (N_2 - N_1)$, which is to say, the heat absorbed is equal to its temperature times the increase in entropy. In the Figure we see that T_1 is the height of the line *ab* above the base line, and $N_2 - N_1$ is its length. Hence their product is the area of the rectangle *abfg*, and this area represents the heat Q_1 absorbed along *ab*. All areas on this diagram represent heat quantities, just as on the *pv* diagram they represented work quantities. The present is therefore a *thermal* diagram, the *pv* is a *mechanical* diagram. Similarly, the heat rejected along *cd* is $Q_2 = T_2 (N_2 - N_1)$, and is represented by the area *cdgf*. The heat turned into work, the difference between these two, is $Q_1 - Q_2 = (T_1 - T_2) (N_2 - N_1)$. It is equal to the difference between the temperatures of the isothermals, multiplied by the difference between the entropies of the adiabatics, and is represented by the

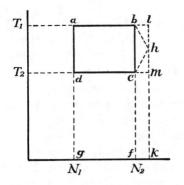

FIG. 22. TEMPERATURE-ENTROPY PLOT OF THE CARNOT CYCLE

area enclosed by the cycle. The efficiency of the cycle is the ratio of its area *abcd* to the area *abfg*, representing Q_1. Thus the efficiency of a cycle shows conspicuously on this diagram, whereas on the *pv* plot it does not show at all.

There are no areas under the vertical lines *ad* and *bc*, for no heat is absorbed or rejected along the adiabatics. Along these lines, heat falls or is raised in temperature without change in its entropy or *heat-weight*. For this reason Zeuner called the adiabatic the *plumb-line* of heat energy.

We may now very accurately compare the action of the cycle of Figure 22 with that of a chain of buckets that runs around pulleys at the four corners. First we may note that the buckets themselves have a certain weight. This corresponds to the entropy N_1 that is already in the system. Along *ab*, water, that is, weight, is added until arrived at *b* a bucket is full and has the total weight N_2. Along *bc*, the buckets are lowered without gain or loss of weight. Along *cd*, the buckets are emptied until at *d* all the water, that is, all the additional weight, has been removed, and the empty buckets with only their own weight N_1 are raised along *da*. The work to be obtained from such a water-motor is equal to the weight added along *ab*, which is the difference between the weights of the full and empty buckets, times the distance of the descent *bc*. This latter may be expressed as the difference between the heights of *ab* and *cd*. Exactly the same formula thus applies, whether we are using water-weight or heat-weight. Heat energy, then, like all other forms of energy, consists of two factors, a capacity and an intensity factor, entropy and temperature—just as gravitational energy depends on weight and height, kinetic energy on mass and velocity, electrical energy on quantity of electricity and difference in potential.

By means of Figure 22 we can also give a simple geometrical proof that the Carnot or rectangular cycle is the most efficient. Suppose that we replace the adiabatic *bc* by the broken line *bhc*, shown dotted. The capacity of the engine is increased, for the area of the triangle *bhc* represents additional heat turned into work. But more than a proportionately greater amount of heat must be drawn from the source, for this additional supply is represented by the area *bhkf*, and it is easy to see that the ratio of the triangle to this area is smaller than the ratio of *blmc* to *blkf* which would have extended the cycle in a perfect manner. In fact we have violated Carnot's rule, for along *bh* the temperature is falling and heat is dropping into the cycle from the temperature level *bl*. Along *hc* the temperature is still falling and heat is dropping out of the cycle before the lower temperature level *mc* is reached. We miss the triangular areas *blh* and *hmc* which might have been turned into work, if all of the heat had been added at the highest and rejected at the lowest temperature. It can be shown that any different line substituted for any other line of the cycle would likewise decrease its efficiency. The only exception is a parallelogram between the same temperature levels. This would have the same efficiency as the rectangle.

These temperature-entropy (TN) diagrams are extremely useful in heat engineering. They show at once where the heat losses occur, and the areas show how much they are. The *pv* indicator card of an engine can always be transformed into a TN diagram, as soon as the more important temperatures are known.

The reader must not be discouraged if he does not yet clearly understand what entropy is. Nobody else did at this point. It turned out to be a useful mathematical function, and most of those who worked with it insisted that it was nothing more—simply a ratio of heat to temperature. They refused to essay any further explanation, and regarded as useless and puerile all attempts to give it a physical meaning that could be commonly understood. Science must retain some mysteries or she will not be so highly respected. Velocity could also be regarded as merely a useful mathematical function, the ratio of distance to time. But because of our familiarity and direct experience with motion, velocity means more to us than a mere ratio. Familiarity with the workings of entropy may also make it seem more real— or at least more important—but we can have no direct experience of it. We cannot feel it like temperature, or see its effects as we can those of heat. Our knowledge of it is necessarily roundabout, and our conception of it will perhaps always be a little vague.

Another difficulty is that there is nothing else like it in the universe, nothing to which it can be accurately compared. All other variables can be increased or diminished. But entropy on the whole always increases. It is a one-way variable. It can only be decreased temporarily and in a restricted region, and then only at the expense of a greater increase elsewhere. The general increase is as inexorable and unpreventable as the march of time. Indeed, it seems to be connected with the march of time, with "the great independent variable." It is the one physical change that unequivocally marks the universe as older to-day than it was yesterday. Entropy is, as Eddington expressed it, "time's arrow."

As age creeps upon us unawares and we scarcely observe the

signs until they have become conspicuous, so we have only recently discovered the real signs of old age in the universe. The
prime sign of old age in a man is a decrease in adaptability. He
becomes set in his ways and ideas; there is less variety in his
thoughts and actions. So the prime sign of old age in the universe is a decreased transformability, a lessened variety, a trend
to uniform monotony.

A new conception of entropy, which gives it a little more
physical meaning, and throws some light upon the wherefore
of its workings, was introduced by Boltzmann in 1877 and has
since been elaborated by Planck and by many others. It is
based on the statistical mechanics, which in turn is an outgrowth
of the kinetic theory of gases. Here we have to deal with enormous numbers of molecules flying about in helter-skelter fashion,
and presenting a perfect picture of primitive chaos. Yet, while
the peregrinations of the individual molecules may be erratic in
the extreme and show no semblance of law and order, the behavior of the crowd as a whole and on the average is very regular.
The gas obeys Boyle's law, its temperature is proportional to the
average kinetic energy of its molecules, it obeys Gay-Lussac's law,
etc. In short, the actions of the individual molecules are like
chance events. While no single event can be predicted, the
general and average behavior can be deduced from the laws of
probability.

These laws apply to any large assemblage of like events, each
of which can happen in several ways, and all of these ways are
equally likely. By *equally likely,* we mean that there is no dis-
cernible reason or evidence why any one of these ways should
occur rather than any other. The theory of probability is hence
a device to circumvent our ignorance. Curiously enough, the
greater the ignorance, the better it works. If any reason can be
shown why the events should happen in one way rather than
in another, then there is a *bias* in that direction—the dice are
loaded—and the laws of probability apply only partially or not at
all.

This does not mean that the laws of probability apply only
to completely indeterminate events—supposing there are such.
The actions of the molecules of a gas are no doubt as fully and
definitely determined as any larger physical events, only, because of the number and complication of the causes at work, we

cannot predict the individual actions. The number π, for example, the ratio of the circumference to the diameter of a circle, has been calculated to 707 decimal places. If this string of figures were the result of mere chance, if, for instance, each had been selected by turning a roulette wheel with the ten digits marked on the circumference, we would expect one-tenth of the numbers to be *sevens,* not exactly, but nearly. Now this is actually so, not exactly, but so nearly that we could not by the application of this or any other probability test decide whether the series was produced by chance or by the operation of some unknown regular law. The only conclusion we could draw is that whatever the process by which the numbers were generated, there was nothing in it that favored one number rather than another. Quite different is it with the decimal value of 1/9, which is .1111 . . . . Here the process is such that no other number can possibly appear. The dice are absolutely loaded, and probability cannot be applied at all.

Now in a gas we do not deal with a measly little number like 707. The number of molecules in one cubic inch of a gas at ordinary temperature and pressure, is 450 billion billion—45 followed by nineteen zeros. Each molecule can move in any direction with any velocity within limits, that is, in an almost infinite number of ways; and each of these can be brought about by the impact of other molecules in an equally numberless variety of ways. There are no restrictions or rules that we can discern as to which of the possibilities will occur at a given moment or place. The laws of probability hence apply very exactly.

Under these circumstances, any semblance of orderliness or regularity is very improbable, and the more improbable the greater the degree, extent, and duration of the orderliness. A large number of molecules moving in the same direction with the same velocity would be extremely improbable; and if by any chance it should occur, it would soon break up again into disorderly movements. Orderly states are therefore rare and improbable. Disorderly states are the rule and are probable. The general trend is from order to disorder, from the less to the more probable states, and finally to a state of complete disorderliness or *chaos,* which is the most probable of all.

Now mechanical energy is due to the regular motions of gross bodies. When such a body is moving, all of its innumerable molecules are moving in the same direction with the same speed. They are like a marching army. They can produce a terrific impact—for the whole kinetic energy of each and every molecule is utilized. After the impact, the molecules become a mob. They rush about in all directions. They still have plenty of energy, but it is no longer fully available. The only way that we can get a part of it, that we can get again a trend in one direction, is to enclose the mob on all sides, leaving a small opening on one side. Then those that chance to be near the opening and going in its general direction will pass out and may even be made to push an obstruction, like a piston, before them. But only a part of the total energy can be obtained in this way. The rest is nullified by the contrary motions, by the cross-purposes of the mob. Never again can we get them *all* to march together with one accord in a single direction. Never again can we utilize their total energy. Thus it is easy to produce disorder, but hard to bring any order out of chaos. It is easy to convert mechanical energy into heat, but hard to convert any of it back again, impossible to reconvert all of it.

The passage from order to disorder, from an improbable to a probable state, is hence also a passage of energy from a more to a less available state, and this is the same as an increase in entropy. Probability and entropy thus vary together. They are connected. Boltzmann, in fact, succeeded in establishing a mathematical relation between them. Entropy is the measure of the disorderliness of a state.

But it takes more than one molecule to create a disorder. Entropy therefore applies only to a crowd. It cannot be applied to a single molecule, or to a gross body moving as a whole—and therefore equivalent to a mass-point. It cannot be applied to mechanical energy, for here there is no disorder. Indeed, all purely mechanical operations are completely reversible. This means that the concerted motions of the huge crowd of molecules that compose a gross moving body cannot have arisen by chance. The improbability is too enormous. They cannot be due, that is to say, to the multiplicity of causes that produce the irregular heat motions. The gross bodies, as wholes, obey the simple laws of mechanics.

Entropy can be applied, however, to the *heat* contained in a gross moving body. Because of the irregular heat motions, the molecules do not move in perfectly straight parallel lines, but in crinkly lines. They march in route step. There is already some disorder in the ranks, and this disorder becomes complete after an impact has arrested the forward march.

Similarly temperature, being the *average* kinetic energy of the moving molecules, can only be applied to a crowd—for it takes more than one molecule to produce an average. Hence all the three thermal quantities, heat, temperature, and entropy apply only to more or less disorderly crowds. They are statistical quantities.

Democritus and his followers held that everything that exists arose from the "fortuitous concourse of the atoms." Given time enough, they reasoned, every possible combination must eventually occur, and so also the present universe. In like vein Thomas Huxley once said, that if six monkeys were set to pounding the keys of six typewriters at random, they would in time write all the books in the British Museum. Perhaps! But in what time? Both of these thinkers vastly underestimated first—the extent of orderliness in the universe, and second—the tremendous improbability of even a slight amount of orderliness arising by mere chance, or long enduring if it did. To write even a single coherent page, the monkeys would require a time so long that the billions of years that we estimate to be the life of a star, would be but the fraction of a second in comparison. Even then the event would not be certain, but only raised to a gambler's chance. It could never be made certain.

Now the books in the British Museum are only one incident in the long orderly life of the universe. This orderliness could not possibly have arisen by chance. Rather we must say that the physical laws that govern the universe leave nothing to chance. Chance cannot bring order out of chaos. Instead, it would destroy any order that existed. The whole trend is *toward* not *away* from chaos.

We resort to probability to circumvent our ignorance. We have no need of this method so long as we confine our attention to a few relatively large bodies. There is no difficulty in following and in accounting for their motions. There is here no entropy, and no increase in entropy. The difficulty begins only when we transfer our attention to the complicated motions of the minute and innumerable molecules. We cannot follow them. They seem confused and aimless, and we note that the general trend is toward ever greater confusion. But the confusion is *ours*. The molecules are never in doubt; *they* never hesitate. Increase in entropy is increase in ignorance.

Everywhere else in physics, it is possible to keep the observer out, to describe the world objectively, as it would go on if no human being ever existed. But in this field, the observer enters as·an unavoidable participant. His limitations are an essential part of the picture. We can discover simple laws and assert a definite determinism only when actions frequently repeat themselves in the same way. We call such actions *regular*. When the actions seldom repeat, or never during the time that we can observe them, we call them *irregular*. An orderly state is one that means something to us, or that we can use. A disorderly state is one that is meaningless or useless. But apart from us human beings, one order is as good as another, A pack of cards, for instance, as it comes from the maker arranged according to suits and number, appears very orderly. If all the inhabitants of the globe shuffled cards steadily for a trillion trillion trillion times the life of a star, there would then be around an even chance of repeating that order once. It would seem therefore to be a very rare and special order. But it is not. Any one of the other twenty trillion raised to the fifth power possible arrangements is just as rare, and would require just as much shuffling to reproduce. The first order *means* something to us. Most of the others do not. Certain other combinations distinguish themselves by their usefulness. Some are useful in bridge, others in poker, etc.

To say that the entropy of the universe tends to a maximum, is simply to say that the universe is passing from an interesting, useful and significant state, to an uninteresting, useless and meaningless state. Apart from us human beings, that is quite indifferent. There is just as much going on in a chaos as in what appears to us as an orderly universe—only it means nothing to us. We cannot make head or tail of it; we can do nothing with it; we have no use for it; we don't like it; and of course we couldn't live in it.

CHAPTER XVIII

HEAT-ENGINES—HOT-AIR AND STEAM

ONE would think that the best way to develop an engine of the highest possible efficiency would be to follow the Carnot cycle as closely as possible, using air as the working substance. This was indeed the main idea of the old hot-air engines. But at once a great obstacle presents itself. The Carnot gas cycle has a very small capacity. In our *pv* diagrams we have drawn true isothermals, but for the sake of clearness have made the adiabatics much steeper than they really are. There is in fact little difference between the two curves, and a Carnot cycle correctly drawn, as in Figure 23, encloses a very small area. This means that only a small amount of work can be obtained from a large range of pressures and volumes. The former requires a strong cylinder, the latter a large one, and both require a heavy engine. It is particularly difficult to make a large cylinder strong. But despite this and other obstacles, the hope of achieving a close approach to the Carnot efficiency led to the expenditure of a great deal of effort on these hot-air engines.

It was of course impractical alternately to heat and cool the same cylinder, as Watt found out in the case of the steam-engine. Hence, in the hot-air engine, the heating was done in one cylinder, the cooling in another, the air being transferred from one to the other at the proper moments. A fire was built directly under the hot cylinder, while the other was surrounded by a jacket of cooling water. Another difficulty now presented itself. The transfer of heat from a solid body to a dry gas and *vice versa* is very slow. Consequently these engines had to be slow speed. Since power is the *rate* of doing work, this meant again that the engine must be large. Finally there was much trouble from the rapid deterioration of the heated surfaces. They

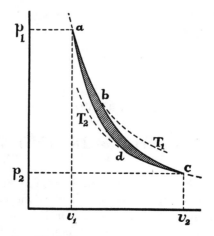

FIG. 23. A Carnot gas cycle to scale

were kept constantly at a dull red heat, and every one knows what happens to a kettle when it is left on the fire after the water has boiled off.

The handicap of small capacity was partly overcome in the engines invented by Stirling in 1827, and by Ericsson in 1833, by the use of the Siemens regenerator. We cannot afford space to describe this apparatus here, and so will only say that even with this improvement, the hot-air engine was still enormous in comparison with a steam-engine of the same power. Yet the fact that these engines theoretically had the Carnot efficiency—the cycle being a parallelogram on the temperature-entropy diagram —and that even the crudest of them actually had an efficiency far exceeding that of the best steam-engines, proved an irresistible lure, and efforts to overcome their various disadvantages continued for many years. Ericsson, inventor of the screw propeller and builder of the Monitor, was the most indefatigable of these workers. In 1854, he equipped a small vessel with his "caloric engine." Although the vessel was only 260 feet long and designed for the very modest speed of seven miles per hour, the engine consisted of four working cylinders 14 feet in diameter, and four compression cylinders 11.4 feet in diameter, the stroke of all being 6 feet. The result was that the boat sank under the weight of its own engines, and with it all hopes of replacing steam by hot air.

Nevertheless, there exist even to-day no engines as efficient as these old Ericsson engines. But, as Carnot himself remarked, efficiency is not the only virtue, or even the most important virtue that an engine can possess. For any sort of a vehicle, smallness of size and lightness of weight are more important.

It is therefore quite impractical to imitate the cycle of Carnot, or his manner of carrying it out. Carnot himself never suggested it. His perfect cycle had only the object of illustrating and developing the theory. His recommendations were to improve the steam-engine by raising the pressure, compounding, and increasing the expansion, and to develop the internal combustion engine. And these are in fact the lines along which real improvements were eventually made.

Every type of engine has a special cycle of its own, which shows how a perfect engine of the sort should work, and would work if there were no thermal or mechanical losses other than those due to its cycle not being perfect. The Carnot cycle shows the maximum efficiency possible between the temperature limits employed, and depends only on those temperatures. The typical cycle gives the maximum efficiency theoretically possible under the added special conditions, and depends also on those conditions. Finally the actual cycle of the engine, as obtained from an indicator card, shows the efficiency attained by the actual engine. The ratio of the typical cycle to the Carnot efficiency measures the excellence of the method employed to convert heat into work. The ratio of the card efficiency to that of the typical cycle shows how well or how poorly the engine carries out that method.

There are at present only three typical cycles in common use— the Rankine, the Otto, and the Diesel, although a few others have been used or proposed. Thus the theoretical and especially the practical possibilities are very limited.

The cycle of the perfect steam-engine is the Rankine. This we have already described in Chapters V and VI, and illustrated in Figures 11, 13, 14. It may surprise some to learn that this cycle of the good old steam-engine more closely approaches the Carnot than that of any other practical engine. On the TN diagram, the cycle is nearly rectangular, consisting as it does of two isothermals, an adiabatic, and a constant volume line. The latter is its only departure from the Carnot. And the indicator card of the engine (Figure 5, Chapter III) closely follows the Rankine cycle, as we also pointed out in Chapter V. The low efficiency of the steam-engine is therefore not due to a poor cycle, nor to a poor carrying out of its cycle, but solely to the narrow temperature limits between which the engine must work. The efficiency of the Rankine cycle may be as high as 95 per cent of that of a Carnot cycle between the same temperature limits, while the further losses in the actual engine may be equally low. There is no other engine in existence so excellent in either of these ways. The steam-engine also has a large capacity, due to the large latent heat of steam. A great deal of heat is carried by the steam into the cylinder at each stroke.

The only way to improve substantially the economy of the steam-engine is to widen its temperature limits, by raising the pressure and increasing the expansion, as Carnot recommended. But here we encounter two practical difficulties, cylinder condensation, and the enormous volume of steam at low pressures, which requires a very large cylinder. The solution of these two difficulties has made our modern highly efficient steam-engines possible.

Watt, it will be remembered, found that the poor economy of the Newcomen engine was due to the condensation of the hot steam when it entered the cold cylinder. His remedy was either to use a nonconducting cylinder, or to keep the cylinder always as hot as the entering steam. But even in a perfectly nonconducting vessel, as we pointed out in Chapter V, steam condenses partially during expansion. Now this condensation, which is inevitable even under perfect conditions, does not affect the theoretical efficiency of the cycle, but it does reduce the practical efficiency of the engine. The entering steam is hot; the exiting steam is cold. The cylinder walls assume a middle temperature. During the first half of expansion, heat passes from the steam to the walls. This heat passes out of the cycle before it has done its full quota of work. During the second half of expansion, heat passes from the walls to the steam. This heat does some work, but not as much as it would have done if put into the cycle at the top temperature. In short, the cylinder walls filch a portion of the heat early in the expansion, hold it out, and drop it in again at a lower temperature, so that the work it would meanwhile have done in the cylinder is lost. If there is a steam jacket, then the walls are always hotter than the expanding steam, and heat is continually dropping in. This heat does some work, but not as much as it would have done if brought in with the entering steam at the top temperature. Now these losses would be small if the steam were always dry, for the interchange of heat between a metal and a dry gas is very slow. But it is greatly aggravated by the presence of moisture, which wets the cylinder and makes a good conducting contact between it and the steam.

The loss can be somewhat reduced by compounding, for then the total temperature drop is divided between two or more cylinders, and the difference in temperature between the steam and the cylinder walls is reduced in each stage. But the most effective remedy is to superheat the steam. This means to heat it above the boiling point corresponding to the pressure employed, and is usually done by carrying the steam after it has left the boiler through a coil of pipe in the smoke-stack. In this way the extra heat costs nothing. Superheated steam is a dry gas, just like any other gas. Its temperature can drop during expansion considerably below the original boiling point before condensation begins, for the condensing temperature drops with the decreasing pressure. If the steam is sufficiently superheated, it may remain dry to the very end of expansion. The use of superheated steam does not appreciably increase the theoretical efficiency of the cycle, but its practical benefit is very great. All engines that pretend to economy now use superheated steam, even locomotives.

The problem of the large volume of steam at low pressures has been met in an entirely different way. We pointed out in Chapter VI that, on account of the working of the Carnot function, more is to be gained by lowering the bottom temperature than by raising the top temperature an equal amount. But to lower the bottom temperature a few degrees means an enormous increase in volume; it means enormously to extend the toe of the Rankine cycle, already very thin. An idea of the relative volumes involved, and of the amount of water produced by various expansions, can be obtained from Figure 24. The circles represent the relative volumes of a pound of water and of the resulting steam at various pressures. They show the relative diameters of cylinders of equal length that would be required to contain the substance, or of pipes that would be required to convey it, provided it always traveled at the same speed. If the speed is increased, the sizes of the pipes can be reduced.

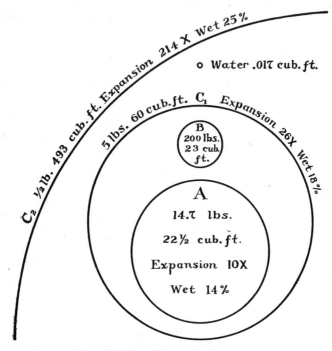

FIG. 24. THE EXPANSION OF STEAM

Compounding is not a solution of this problem, because the low pressure cylinder must in any case be large enough to contain the whole of the expanded steam. Starting with a boiler pressure of 200 pounds absolute, represented by the circle B, and expanding to atmospheric pressure, circle A, the expansion ratio is about 10. This can easily be accommodated, and so can the expansion of 26 times to 5 pounds absolute, represented by C_1. But to expand to $\frac{1}{2}$ pound, represented by the circle C_2, only a quarter of which is shown, would require a cylinder 7 feet in diameter and of $12\frac{1}{2}$ foot stroke to contain one pound of steam. That would be a monstrosity. Besides, the steam is then 25 per cent wet, and it would be impossible to expel this mixture through ports of any reasonable size. Yet 50 per cent more power would be obtained by the additional expansion from 5 pounds to ½ pound.

The steam turbine has made this great expansion possible. Despite its very different mechanism, the turbine works on the same Rankine cycle as does the reciprocating engine. Steam is admitted, expanded, exhausted, and condensed in the same way. But the expansion takes place, either in flaring nozzles, or in widening passages, as the steam threads its way between innumerable pairs of blades that run away from each other. Since the steam is always moving, a space large enough to contain any great quantity of it at one time has to be provided. And its velocity is very high, from 1400 to 4000 feet per second. As the passage widens the velocity of the steam increases, so that the passage does not have to enlarge to the same extent that the steam expands. In fact, a nozzle 1 inch in diameter at the throat and 7 inches at the mouth, will produce the whole expansion from 200 pounds down to ½ pound. The turbine can therefore take care of an enormous expansion in a comparatively small and light machine.

For this reason the low pressure turbine is particularly valuable. Many years ago the Consolidated Edison Company of New York doubled the capacity of one of its plants, consisting of already highly efficient triple expansion engines, by merely adding low pressure turbines which utilized a few pounds pressure drop in the exhaust steam from these engines. At low pressures the turbine thrives where the reciprocating engine cannot exist at all.

Turbines are of two types, impulse and pressure. In the impulse type, the whole expansion takes place in nozzles. These constitute the whole heat-engine. Here the available heat of the steam is converted into kinetic energy. The rest of the turbine is merely a bladed wheel that is driven by the moving steam, just as it would be by jets of water or of air. The steam when it impinges on the blades is already at the exhaust pressure and temperature. No further expansion or cooling takes place in passing between them, and the pressure is the same on both sides of the blades.

The simple impulse wheel has a serious disadvantage. It requires an enormous speed. For maximum efficiency, the speed of the blades must be half that of the impinging steam. Under these circumstances, since the cup-shaped blades reverse the direction of flow, the steam on leaving the blades is at rest, or dead. The wheel has utilized, or taken off, the whole velocity. There must of course be some lateral motion of the steam to carry it away from the wheel, and this is obtained by making the steam impinge upon the blades obliquely, as shown in Figure 25. This lateral velocity is completely lost as far as power production is concerned. Since the velocity of the steam is very high, a correspondingly high speed is required of the wheel. The early De Laval turbine (1889) ran at 30,000 revolutions per minute. It had to be geared down, usually in the ratio of ten to one, even to run a dynamo, for which reciprocating engines usually had to

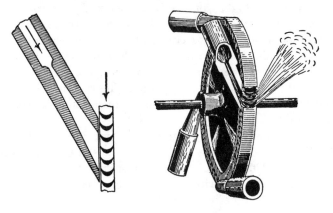

FIG. 25. THE DE LAVAL IMPULSE WHEEL

be geared up. Because of its high speed, however, it was extraordinarily light and compact.

To avoid this excessive speed, turbines have been made having several alternate rows of moving and fixed blades. The latter have simply the function of reversing the flow of the steam after it has left one row of moving blades, so that it will impinge upon the next moving row in the right direction. The steam thus zigzags it way through the turbine. The first row of moving blades takes off a part of its velocity, the second row another part, and so on, until after passing the last row the steam is dead. This is called velocity compounding. Another method was to have several sets of nozzles, each working on a set of blades enclosed in a separate chamber. The first nozzles expanded the steam partially, developing a part of the possible velocity, and this was completely utilized by the one or more rows of blades in the first chamber. The second set of nozzles then expanded the steam further, and the velocity developed by them was utilized by the blades in the second chamber, and so on. In each succeeding chamber a lower pressure prevailed, so that each constituted a stage of expansion. This method was therefore called pressure compounding. By either or both of these methods combined, the economical speed of the turbine could be much reduced, but at the expense of greatly increased size, weight, cost, and complications. These types are now practically all obsolete. They have given way to the pressure turbine invented by Parsons in 1884. Although this machine had its inception almost as early as the impulse turbine, it was much longer in getting into use. But when its superiority was once clearly demonstrated, many almost new impulse turbines were scrapped in its favor.

In the pressure turbine, expanding nozzles are entirely dispensed with. Expansion takes place between the blades themselves, both fixed and moving, which are shaped for that purpose. Each set of blades thus constitutes a stage of expansion, so that instead of the three or four stages, which is the practical limit of the reciprocating engine or of the impulse turbine, we often have in the Parsons turbine a hundred or more. That is multiple expansion indeed. The many sets of moving blades are carried on a long drum or cylinder, the rotor.

Admission of steam to the turbine is through the first set of fixed blades. These are like so many tiny nozzles, and they cover the whole circumference of the rotor. This secures *full admission,* which is impossible with nozzles even though several are used. It increases the capacity of the engine.

The velocity developed by the steam at each stage of expansion is immediately taken off by the next set of moving blades. Hence the velocity can be kept down to a moderate amount, and these turbines can run efficiently at a much lower speed than can any impulse turbine. This has made it possible to apply them to marine propulsion, where low speeds are necessary, a thing impossible to the impulse turbine without gearing.

In the Parsons turbine, there is a difference in pressure between the front and the back of the blade. It is partly this that drives the machine, but impulse and reaction each play a part also. Because of this difference in pressure, there is always some leakage of steam over the end of the blade. But since the pressure drop at each expansion stage is small—because of their great number—this leakage is not serious.

Superheated steam is beneficial in turbines as well as in piston engines, but for quite a different reason. There are no great variations in temperature in any part of a turbine, which cause an interchange of heat between the metal and the steam. When a turbine is once warmed up, there is a continuous drop in temperature from the hot end to the cold end, so that the steam always comes in contact with metal at about its own temperature. But one of the chief losses in a turbine is due to friction between the steam and the blades, and this is greatly increased by moisture, which condenses in fine droplets on the blades and converts an otherwise smooth surface into a rough one.

The steam turbine has been brought today to a high state of perfection. For smoothness, quietness, and evenness of running, for fineness of regulation, wide range of power and speed, ability to take an overload, good efficiency at reduced loads, simplicity, durability, and reliability, it is not surpassed by any other prime-mover. Anything that will burn can be used as fuel. There are no noxious fumes in the exhaust. For large and continuous power production, especially in stationary plants where space and weight are of no great concern, it is still the best and cheapest of the fuel burning engines. Despite hydro-electric plants, gas and oil engines, by far the greater part of the world's billion horsepower is still produced by steam, and most of this by the combustion of fossil fuels.

CHAPTER XIX

THE INTERNAL COMBUSTION ENGINE

HOWEVER efficient the steam-engine may be within its own temperature limits, those limits are necessarily narrow. The great loss of motive power resulting from the free fall in temperature between the fire and the boiler is unavoidable in this type of engine. The only remedy, as Carnot pointed out, is to burn the fuel directly in the cylinder, in short, to develop the internal combustion engine. Experiments with such engines were already in progress in his day, but they were greatly hampered by the high cost of gas or of any other fuel that could be used. Around 1860, explosion engines began to appear in practice. They were very inefficient because they did not compress the gas before igniting it, although Carnot had pointed out the advantage of so doing.

Among these early gas-engines was the curious free-piston engine. This was built on exactly the plan suggested by Huygens in the seventeenth century, which we described on page 5. A heavy piston was shot upward by an explosion beneath it in a tall vertical cylinder. Its upward motion was perfectly free, but in descending, a toothed rack carried by the piston engaged a small gear which turned the engine, while a pawl and ratchet mechanism like the stem-winder of a watch, prevented the engine from being turned backwards during the upstroke. Since no initial compression was used, these engines were inefficient. Also they were insufferably noisy.

In 1876, Otto brought out his "silent engine." It employed a cycle that had been previously described by Beau de Rochas, but which became generally known as the Otto cycle, perhaps because the name is easier to pronounce. It was the first engine to use initial compression. But this important feature, to which it owed its higher efficiency, was little appreciated at the time. Its chief selling point was that it was less noisy than its predecessors.

The theoretical cycle of the Otto engine is shown in Figure 26. On the first forward stroke of the piston, air and fuel, already mixed in the proper proportions, are sucked into the cylinder. This operation is represented by *ab*, a constant pressure atmospheric line. Along *be* the mixture is compressed adiabatically. It will be noticed that this engine has a very large clearance as compared with the steam-engine. The whole cycle is far to the right of the left-hand axis. This is necessary in order to limit the compression. At *c* the mixture is exploded, and the pressure rises abruptly along *cd*. This is equivalent to simply heating the air charge. All gases are much alike. Very little change in physical properties is produced by the chemical change of combustion. Hence we may consider the working substance to be simply air, which is alternately heated and cooled. The combustion line is a constant volume line, for theoretically the explosion is instantaneous and occurs when the engine is on dead center, and the piston stationary at the end of its stroke. Along *de*, the next forward stroke, adiabatic expansion takes place. At *e* the exhaust valve opens, and the pressure drops abruptly along the constant volume line *eb*, the engine being now on the other dead center. Along *ba* the burnt gases are exhausted, or as much of them as

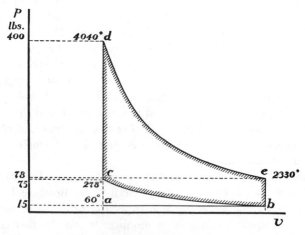

FIG. 26. THE OTTO GAS CYCLE

the length of this line represents. What remains in the clearance space—combustion chamber—is not exhausted, and this is a defect of the engine.

Four strokes are thus required to complete the cycle, for which reason it is often called the four-stroke cycle. But only two of them belong to the real heat cycle, the compression and the power strokes *bc* and *de*. The two strokes *ab* and *ba* are displacement strokes, which merely shift the location of the working substance —a purely mechanical matter—without absorption or expenditure either of heat or power. It is, of course, entirely a misnomer to call the engine a four-cycle engine. There is only one cycle to every engine.

The real heat cycle is then *cdeb*. It consists of two constant volume lines and two adiabatics, for which reason it is also called the constant volume cycle. Heat is received along *cd* and rejected along *eb*. Only instead of the working substance being cooled along *eb* by contact with a cold body, it is thrown bodily out with its contained heat, and a fresh charge already cool is drawn into the engine. This substitution makes no difference in the theoretical cycle. It is merely necessitated by the practical circumstance that, due to the combustion, it is impossible to restore the working substance to the initial condition at *c* by a mere adiabatic compression.

The pressures and temperatures marked on the diagram represent a typical case. The charge is drawn in at 60 degrees Fahrenheit and 15 pounds (atmospheric) pressure. It is then compressed to 75 pounds, a compression ratio of 5. This raises the temperature to 278 degrees. The explosion raises the pressure to 400 pounds, and the temperature to 4040 degrees. During expansion the pressure sinks to 78 pounds at *e*, and the temperature to 2330 degrees. From these data, the heat absorbed, the quantity turned into work, and the efficiency of the cycle can be calculated. For this particular case the cycle efficiency is 38 per cent. The efficiency of a Carnot cycle between the same temperature limits, 4040 and 60 degrees, would be 76 per cent. This Otto cycle thus utilizes only half of the available fall in temperature, and is in this respect much inferior to the Rankine steam cycle, which we saw may utilize as much as 95 per cent. The chief reasons for this poor showing are, first—that the heat is not absorbed and rejected isothermally at the highest and lowest temperatures respectively, but at rising and falling temperatures along the constant volume lines *cd* and *eb;* second—the high temperature and pressure still prevailing at the end of expansion at *e,* represent a considerable amount of work still contained in the working substance when it is thrown out. Finally, the actual engine realizes only about half the efficiency of its typical cycle. The chief reason for this is that, whereas in the steam-engine we do everything possible to keep the cylinder hot, in the gas-engine we have to do everything possible to keep it cool, and the water jacket we apply for this purpose steals from 35 to 50 per cent of the heat supplied. Yet if we did not use it, the temperature attained would soon burn up not only the lubricating oil but the metal as well, as every autoist knows who has let his radiator run dry. Nevertheless for economy the engine should be run as hot as can safely be done.

Despite all these defects, the Otto is still much more efficient than the steam-engine, and this is due solely to the much higher top temperature employed—the temperature of the fire itself. And since the engine is the whole power plant, the boiler losses, space, and weight are saved. If the water jacket steals 35 to 50 per cent of the heat supplied, we must remember that in a steam plant, at least this amount goes up the chimney.

Figure 27 shows what the actual cycle or indicator card of the engine looks like, as compared with its theoretical cycle. All corners are rounded and the area is much reduced. The combustion line *cd* is curved, for the explosion is *not* instantaneous. It starts at *c* before the piston has reached the end of its stroke (the spark being advanced) and ends at *d* after the piston has begun its next forward stroke. Partly because of the increase in volume during combustion, and partly because of the heat stolen by the water jacket, the height of *cd* is much reduced. The pressure does not rise nearly so high, and the temperature only about half as high as in the theoretical cycle. However, even this temperature of around 2000 degrees would be fatal if it were sustained. But it only lasts a small fraction of the time occupied by the four strokes, which themselves take up but a fraction of a second.

The efficiency of the ideal Otto cycle depends, as can be shown, solely on the ratio of compression, and this is also the chief determining factor of the practical efficiency. The compression

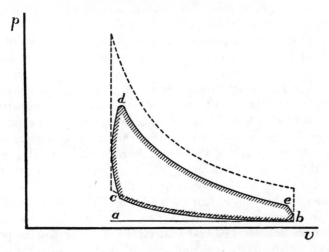

FIG. 27. INDICATOR CARD OF A GAS-ENGINE

should therefore be made as high as possible. But since to raise the pressure raises also the temperature at the end of compression, danger of pre-ignition arises. There is hence a limit to

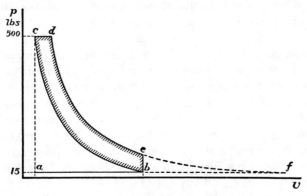

FIG. 28. THE DIESEL CYCLE

improvement in this direction, and modern engines have about reached that limit.

This evil of pre-ignition, which limits the efficiency of the Otto engine, was turned into a virtue in a cycle and engine invented by Rudolf Diesel in 1897. The theoretical cycle of this engine is shown in Figure 28. Pure air is drawn in along *ab,* and is compressed along *bc* to about 500 pounds—a compression ratio of 30 to 35 being used in this engine as against 5 to 6 in the Otto. This raises the temperature above the ignition point of the fuel, which being then injected by an atomizer, takes fire at once. All ignition systems are thus dispensed with. The fuel injection and burning continue along *cd* at constant pressure and rising temperature, as the piston moves out on its power stroke. The point *d* is like the cut-off of a steam-engine. The fuel injection is there cut off, and the gas then expands adiabatically along *de* for the rest of the stroke. At *e* the exhaust valve opens and the pressure drops along *eb*. The burned gases are then expelled along *ba*. It may be noted that the clearance is much smaller in the Diesel than in the Otto. This is because of the higher compression. The burned gases are therefore more completely exhausted. The real heat cycle is *cdeb* and is composed of two adiabatics, a constant pressure, and a constant volume line. Heat is taken in at constant pressure along *cd,* and is rejected at constant volume along *eb*.

The Diesel cycle is hence much like the Otto except for the higher compression and the prolonged combustion. Its efficiency depends mainly on the ratio of compression, but also on the position of the cut-off, decreasing as the latter is delayed. The efficiency is higher than the Otto, mainly because of the higher compression. The engine, however, is heavier, for the space between the two long adiabatics is rather narrow, meaning little work from a large cylinder. The running is not quite so jerky as that of the Otto, because of the prolonged combustion, the line *cd* resembling the admission line of the steam cycle. But the running is far from as smooth as that of a steam-engine, for the cut-off is usually at about one-tenth of the stroke, and the pressure drop from 500 to 15 pounds is much larger than is used in any single expansion steam-engine.

If in the Diesel cycle we prolong the expansion line *de* down to atmospheric pressure at *f,* we obtain a cycle *cdfb,* composed of two adiabatics and the two constant pressure lines *cd* and *fb*. This is called the constant pressure cycle. It was first proposed by Joule, and so is also called the Joule cycle. Evidently the Diesel is a transition form between the constant volume Otto, and this constant pressure cycle. The latter is even more efficient than the Diesel, for it utilizes the available work thrown away in the exhaust, and represented by the area of the dotted toe. This cycle has been experimented with, but so far without much practical success. To carry it out in an engine of the Diesel type would require strokes of two different lengths, two short ones, *ab* and *bc,* and two long ones, *cdef* and *fa.* It is easy enough to devise a mechanism that will accomplish this, but a complicated mechanism right in the part of an engine that has to endure the greatest strains and temperature variations, has been found to be impractical. The Diesel is already a complicated engine, and every added complication is another liability. There are, however, other and quite different methods of carrying out the Joule cycle, one of which will be discussed presently. Some engineers think it will yet be the cycle of the future.

Much labor, time, money, and ingenuity have been wasted in attempts to produce a *rotary* gas engine, that is, one in which only rotating motions occur. Since in every case the cycle involved is simply the Otto or the Diesel, there can be no gain in thermal efficiency, and the only possible improvements are purely mechanical. The main idea has been to get rid of reciprocating parts. But this is not nearly so important as has been imagined. Improvements in the mechanical design and construction of existing engines, their proper balancing and compensating, on which an immense amount of research and engineering skill have been expended, have so far reduced vibration and other inconveniences of reciprocating parts, that their total elimination would mean but little gain even mechanically. But the rotary engine does not completely accomplish this elimination. The alternately contracting and expanding spaces required to carry out the gas cycle are in the main produced by rotating parts, but usually some sliding pieces are also required, or at least pieces which oscillate in addition to rotating. Hence the problem of balancing and compensating has to be solved all over again, and under much more difficult conditions.

Besides, these engines have difficulties of their own. All or a large part of the complicated mechanism is right in the fire of the engine, in the combustion chamber, hence subject to rapid deterioration or to overheating, and difficult to lubricate. Repairs can only be made by taking the engine completely apart. Frequently there is no adequate provision to prevent leakage from the high to the low pressure compartments. Hence both mechanically and thermally these engines have been inferior to the piston engine. There is only one possible advantage, and that is high capacity. That should make them suitable for vehicles provided, of course, the other difficulties can be overcome.

A more promising possibility is the gas turbine. This really does get rid of all reciprocating, sliding, or oscillating parts, for there is only one moving part—the rotor, and its motion is *uniform* rotation. Yet this is not its chief advantage. Its chief advantage is thermal rather than mechanical, in that, like the steam turbine, it enables us to develop the toe of the indicator card, and thus increase *efficiency* as well as power. The gas turbine can do this for either the Otto or the Diesel cycle, and the latter is then converted into the Joule, the most efficient gas cycle known except the Carnot. The simple gas turbine power plant consists of three major elements: air compressor, combustion changer and turbine. A starting motor, a generator, a governing system and a lubrication system are also required. Gas turbines have been widely used in airplanes, for stationary power generation, for pipe lines and to drive transportation equipment ranging from pleasure cars to locomotives and ships.

CHAPTER XX

MECHANICAL REFRIGERATION

THE MECHANICAL refrigerator is a heat-pump. By means of a reversed cycle, it lifts heat from the body to be cooled to a temperature somewhat above that of the surroundings, and allows it to flow away. Theoretically any heat cycle would do, but practically only two have been used—a modified Rankine for vapors, and the Joule cycle for the air.

The idea of the air cooler was first proposed by William Thomson in 1852, before the energy theory had scarcely emerged from the cradle, but actual machines did not appear until two decades later. The reversed Joule cycle on which they work is shown in Figure 29. Starting at *b*, a charge of air at 15 pounds pressure and 60 degrees Fahrenheit is compressed adiabatically along *bc* to say 60 pounds. This raises its temperature to 313 degrees. The air, while still under this pressure, is then passed through a coil of pipe immersed in running water by which it is cooled to the original temperature. This operation is represented by *ce* and takes place while the air is passed from the compression cylinder to another and smaller expansion cylinder. Here the adiabatic expansion *ef* takes place, the temperature meanwhile dropping to −111 degrees—under the conditions assumed. The cold air is then exhausted into the room to be cooled. The pistons of the two cylinders work on the same shaft, so that the work done by the expansion is recovered. The work of compression is given by the area *abcd*, that of expansion by *defa*, the net work put into the cycle by the enclosed area *fbce*. The actual imperfect machine, however, did not attain anything like the temperature range or the efficiency of this theoretical diagram.

Machines of this sort began to appear commercially in 1873.

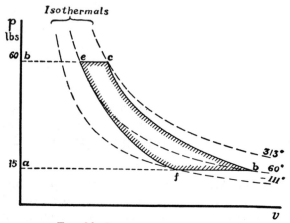

FIG. 29. CYCLE OF THE AIR COOLER

They were much used on ship board, and continued to be so used even after the appearance of the ammonia machines, because of the absence of all danger of noxious fumes. They were, like the hot-air engines, bulky, heavy, and attained only a fraction of their theoretical efficiencies. Much trouble was experienced also from moisture in the air, which would condense to snow, and was difficult to exhaust from the machines. They have been replaced on ship board largely by carbon dioxide machines, which work on the same principle as the ammonia machine.

Vapor compression machines were experimented with as early as 1834 (Jacob Perkins of Boston), but the modern ammonia compressor was invented by Carl Linde of Munich in 1876. It works on a reversed steam cycle. Steam of course cannot be used, because water is solid below 32 degrees and boils at 212. Liquid ammonia (not to be confused with the household article, which is a solution of ammonia gas in water) boils at −28 degrees, and is consequently gaseous at all higher temperatures, unless a considerable pressure is applied. Various other liquids of low boiling point have also been used.

The process consists essentially in allowing the liquid ammonia to boil at a low temperature by abstracting the required latent heat from the bodies to be cooled. The resulting vapor is then compressed, condensed by cooling water, and cooled to room temperature. The liquid ammonia is then further cooled by expansion, and again boiled at the low temperature. Thus we boil the liquid at a low temperature and pressure, and condense the vapor at a high temperature and pressure, the reverse of the operations of a steam engine.

The arrangement of an ammonia compression plant is shown in Figure 30. Liquid ammonia, a supply of which is kept in the reservoir at the left under about 200 pounds pressure and at room temperature, passes down to the needle valve at V. Here it is sprayed into a coil of larger pipe immersed in a brine tank. The pressure in this pipe is about 30 pounds. The sudden drop in pressure causes some of the ammonia to evaporate and to lower its temperature to 0 degree Fahrenheit, its boiling point at

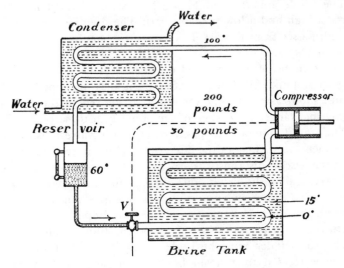

FIG. 30. AN AMMONIA COMPRESSION PLANT

30 pounds pressure. The temperature of the brine is about 15 degrees, so that heat flows from it to the ammonia, supplying the latent heat necessary to boil the latter. Thus the brine tank is our low temperature boiler. The vapor is then compressed by the compressor on the right to 200 pounds, which raises its temperature to 100 degrees. It then passes through the condenser where it is liquefied and cooled to room temperature, and finally collected in the reservoir, ready for another round of the

cycle. The liquid ammonia in this condition is called the refrigerant. It can be kept indefinitely and even transported in sealed containers, without losing its refrigerating power, for, being at the temperature of the surroundings, it cannot absorb heat. The refrigeration can always be obtained by merely releasing the pressure. The same is true of any vapor kept liquid at ordinary temperature by pressure. It is even true of a compressed gas. But a liquid refrigerant is much more powerful than a gaseous one, because in vaporizing it must absorb its latent heat, always a large quantity, while the latter can only absorb its specific heat, always a small quantity. Besides, a much greater weight of liquid than of gas can be stored in a given container at a given pressure.

The whole apparatus of Figure 30 may be divided into two parts, a high and a low pressure part. The dividing line, shown dotted, passes from the needle valve to the compressor, the two pressure changing organs.

The cold brine may be used directly to make ice, by immersing in it tanks filled with water, or it may be piped to the rooms, show-cases, or whatever else it is desired to cool. There is then no danger of leakage of ammonia into those places.

The liquid ammonia in passing the needle valve must, as we have mentioned, first refrigerate itself before it can refrigerate anything else. This constitutes a loss, which might be avoided by carrying out the expansion adiabatically in a cylinder provided with a piston, by which the work of expansion could be recovered. But the gain at most would be only 10 per cent. It is not worth the extra complication. In short, the liquid ammonia fed to the cold boiler must be cooled to boiler temperature, just as the water fed to a steam boiler must be heated to boiler temperature. We saw that this heating could have been done more perfectly by compressing the water adiabatically—mixed with a little steam that is condensed during the operation—but that the gain was not worth the extra complication.

The ammonia compressor is superior to the air-cooler for much the same reasons that the steam-engine is superior to the old hot-air engines. It is much more compact, and this superior capacity is due to the large latent heat of the ammonia, 566 therms per pound, which must be absorbed before its temperature can rise above 0 degree. It is due also to the more rapid interchange of heat between a vapor or a liquid and its surroundings, than between a gas and its surroundings. The ammonia process also realizes a larger percentage of its Carnot efficiency, just as the steam-engine does, because both the absorption and rejection of heat are largely isothermal.

The electric refrigerators that are so necessary in our homes work in just this same way. But instead of ammonia, some of them use sulphur dioxide, because of the less pressure required—80 instead of 200 pounds. Other refrigerants, like the Freons, are also used. The brine is dispensed with, the coil of pipe into which the refrigerant expands doing the cooling directly, and in place of the condenser and cooling water, an air-cooled radiator is used, assisted by a fan. If one puts his hand on this radiator, he can feel the heat that has been lifted out of the refrigerator.

There is another and older method of ammonia refrigeration called the absorption process. The cycle is the same, but the pressure is produced, not by applying power, but by heating a solution of ammonia gas in water—the household variety. Ammonia is extraordinarily soluble in water, one volume of water at 60 degrees taking up 800 volumes of the gas. The solubility also varies greatly with the temperature, reaching 1300 volumes at 32 degrees. Hence if a solution saturated at 60 degrees is heated, immense volumes of the gas are driven off. If heated in a closed container, a pressure is produced as in a steam boiler.

A simple form of apparatus applying this principle was invented by Carré in 1860. It consists merely of two bulbs connected by a tube as shown in Figure 31. The bulb A in the upper figure contains a *strong* solution of ammonia. When a flame is applied, the gas is driven off and a pressure is produced. The vapor passes into the other bulb B immersed in cold water, where the vapor is condensed. We now have our liquid refrigerant under pressure. When enough has been collected, the flame is removed from A and a cold bath substituted for it, as shown in the lower Figure. At the reduced temperature, the now *weak* solution in A reabsorbs the vapor, producing a partial vacuum in B which causes the liquid there to vaporize, and draw heat from the brine or whatever else the bulb may now be immersed in. The two operations are repeated as many times as are necessary to produce the desired refrigeration.

For commercial purposes it was necessary to make the operation continuous. Each bulb in the above apparatus, it will be
noted, performs two functions. First, A and B are generator and
condenser respectively; then they become absorber and evaporator. The first step was to separate these functions, and to pro-

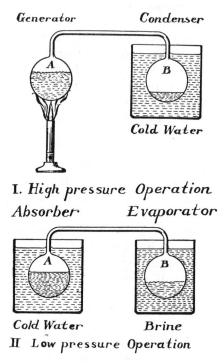

I. *High pressure Operation*

II *Low pressure Operation*

FIG. 31. SIMPLE AMMONIA ABSORPTION REFRIGERATOR

vide a separate organ for each. The lay-out of a commercial absorption plant is shown in Figure 32. It is exactly like that of a
compressor plant, except that the compressor is replaced by the
four pieces of apparatus at the right. The operation is also much
the same. The liquid ammonia, after passing the needle valve V,
vaporizes in the evaporator, and the gas passes into the absorber.
The latter contains a weak solution of ammonia, which absorbs
the gas. The generator above it contains a strong solution of
ammonia, and this being heated by the fire beneath, the gas is
driven off and a pressure is created. The gas, as before, passes

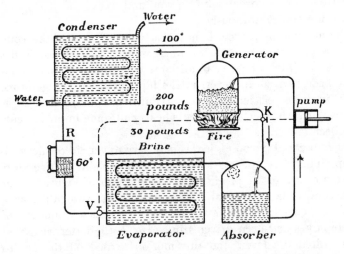

FIG. 32. COMMERCIAL AMMONIA ABSORPTION PLANT

through the condenser, where it is liquefied and cooled, and the refrigerant is collected in the reservoir on the left. This process could not go on very long, because the solution in the generator is continually growing weaker, that in the absorber stronger. The solutions have to be interchanged. Now in the generator, the solution is weaker near the bottom. This weaker solution is drawn off, passes through the pressure reducing valve K (for the generator is in the 200 pound region), and runs down into the top of the absorber. Here the conditions are reversed. The stronger solution is at the bottom. This stronger part is drawn off, pumped up-pressure by the small feed-pump on the right, and delivered to the top of the small generator. We have also to reckon with the fact that the generator is hot, the absorber cold; and this interchange of solutions tends to equalize the temperatures. To prevent this, a heat interchanger is interposed between the two vessels. To avoid confusion, this apparatus is not shown in the diagram. In it ascending and descending pipes are coiled together or made one concentric with the other, so that the cold ascending solution is heated by the hot descending solution, and *vice versa,* the latter is cooled by the cold ascending solution. By these means the generator solution is kept hot and strong, the absorber solution cold and weak, and the operations can go on continuously.

The absorption process is not nearly so efficient as the compression process, and requires much more space. Hence it has been replaced by the latter in all large commercial plants. It has lately been revived, however, in the household gas refrigerators. Here, its lower efficiency is not of much moment, because gas heat is much cheaper than electric heat, and some improvements have been made.

For household purposes, it was necessary to get rid of the feed-pump. This was not easy. The early household refrigerators staged a return to the primitive. The apparatus of Figure 31 was used. It was made automatic by balancing the two bulbs by a fulcrum between them, like a lever. When a sufficient amount of ammonia gas had been driven from A to B, the latter, weighted down, tilted the lever, and this motion turned off the gas to the burner. The mere coolness of the surroundings in the absence of the flame was then relied upon to cool A enough to reduce the pressure, so that the liquid in B reëvaporated and passed back into A. When the latter was sufficiently over-weighted, the lever tilted back again and turned on the gas to the burner, which was ignited by a small pilot kept always burning. Thus tilting back and forth, the brine surrounding B was gradually cooled below the freezing point. But the efficiency of this apparatus was too low even for household purposes.

The modern gas refrigerator has solved the pump problem in a very ingenious manner. The plant is substantially the same as Figure 32, except that the pump and reducing valve are omitted, and a special form of solution and heat interchanger is substituted. Hydrogen gas at a pressure of 170 pounds is introduced into the low pressure part of the apparatus—the evaporator and the absorber—and this with the 30 pounds pressure of the ammonia vapor there present, brings the total pressure of this part up to 200 pounds, the same as prevails in the high pressure part. This is simply an application of Dalton's law of partial pressures. The ammonia evaporates into this hydrogen atmosphere almost as readily as though nothing were there. (Several gases were tried, but hydrogen was found to be the best.) Furthermore, the hydrogen is made to circulate from the evaporator to the absorber and back again, and thus constantly sweeping over the evaporating ammonia actually promotes the evaporation of the latter. There being then the same total pressure in all parts of the apparatus, the interchange of the solutions is accomplished by simple thermosiphon action. The absorber is placed above the generator, and the heavy strong solution at the bottom of it runs down by gravity to the generator. The heat applied to the latter is made to raise the weaker part of its solution to the top of the absorber. Thus the operations go on continuously and silently without the use of any moving parts.

The usual way of heating a house is to burn fuel. If we do this in a fireplace, 95 per cent of the heat goes up the chimney. If we use a stove or a furnace we do not lose so much, but still a good deal. If we burn a gas or electric stove in the room without a flue, we lose none of the heat. Such a heater is 100 per cent efficient. More than that would seem impossible. Hence if a man told you that he had an apparatus by which he could obtain four times as much heat as was developed by the fire, a heater in short that was 400 per cent efficient, you would doubtless take him for some sort of a perpetual motion crank. If in addition he told you this his apparatus was a refrigerator, you would then *know* he was an escaped lunatic. Yet that is precisely what Thomson proposed in 1852, long before the first refrigerating machine was built.[1]

Consider what a refrigerating machine does. It lifts heat from a lower to a higher temperature. Ordinarily it lifts heat from a temperature much below the surroundings to one slightly above, and then throws it away. Why not put the radiator of our refrigerator in the living room and utilize this heat, instead of leaving it to warm up the back porch, or the kitchen which is already warm? Or better still, instead of lifting heat from a small body, let us lift it from the great outdoors where there is plenty of it, and deliver it to the house. This of course requires power. But when we build a fire to warm a house, we allow the temperature to drop from that of the fire, some 2000 degrees, to that of the room, 70 degrees. We could utilize this drop to develop mechanical power, and still have *all* of the heat on

[1] "On the economy of the heating or cooling of buildings by means of currents of air," *Proceedings of the Glasgow Philosophical Society*, Vol. III, p. 269.

hand. The mechanical power we could use to run a refrigerating machine that would lift a much larger quantity of heat through the much smaller difference of temperature between the outdoors and the indoors, and thus get four or five times as much heat into the house as was developed by the fire. There is no perpetual motion or conflict with the conservation of energy in this, any more than there is in the fact that a small weight on the long arm of a lever descending three feet, can raise a weight three times as great on the short arm to a height of one foot.

The situation is analogous to that of a man who has a house on a river bank ten feet above the stream. A hundred feet higher up on the bank is a small reservoir. The man wants water to irrigate his garden. He could, of course, simply let the water run down from the reservoir. But suppose that is not enough. Now the water in descending a hundred feet to the garden can develop mechanical power. Let the man use this power to run a pump. Assuming 100 per cent efficiency, the pump could raise ten times as much water up ten feet from the river, as descended from the reservoir. He would get in all eleven times as much water into his garden as was contained in the reservoir. This is precisely the action of the heat pump.

Indeed, it seems a crime to let heat drop from 2000 to 70 degrees without developing any of the mechanical power that this drop makes available. But many practical difficulties have prevented any wide use of the heat pump—high first cost, complicated machinery, expert attendance, and above all, slow heating. This last is due to the very moderate temperature at which the heat must be supplied to the house, and this requires large radiators. We are in too much of a hurry. On a cold morning, we want to heat the house quickly, and only high temperature heat can do this. In a large part of Europe, where fuel is expensive, huge porcelain stoves have already long made use of low temperature heat. The heat pump has been applied to a few large buildings in this country and will undoubtedly come more into use as our fuel resources diminish. For small houses, the gas refrigerator might be applied. There being no engine, no expert attendance is required. One could turn on the gas, or build a small fire, and get four or five times as much heat into the house as the fire developed.

When it is desired to reach very low temperatures, such as are required for the liquefaction of gases, other means must be employed than those we have described. The method by which liquid air is to-day produced in commercial quantities is based on the cooling of a gas on free expansion, discovered by Thomson and Joule in 1848. This is combined with the Siemens regenerator or heat interchanger. The process was perfected by Carl Linde in 1893, who has thus given us both of our two most important methods of mechanical refrigeration. An early form of his machine is shown in Figure 33. Air is highly compressed by a

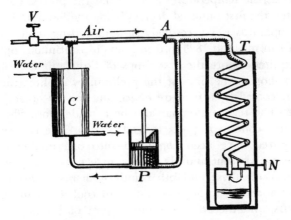

FIG. 33. LINDE'S LIQUID AIR MACHINE

powerful compressor at P, passes through condenser C, where the heat of compression is removed, and is then conveyed by the small pipe at the top, which at A enters a larger pipe which surrounds it concentrically. The two are formed into a coil in the container T, which is packed with a great quantity of heat insulating material. At the bottom of the coil, the smaller pipe is brought out, and the compressed air is expanded through the needle valve N to a lower pressure. This is a free expansion, and there is a small drop in temperature. The cooled air then passes back through the large pipe that surrounds the small one, and thus cools the air that is approaching the needle valve. This is the regenerator principle. The expanded air continues on up out of T, down the large vertical pipe, and back to the compressor, where it is recompressed and sent around again. Each time the air passes the needle valve, a further drop in temperature occurs, so that as the process continues, the temperature at N gradually descends, until at last liquid air drips from the valve to the container below. As the air is liquefied and drawn off, new air must be pumped into the apparatus to replace it. This is done by another compressor not shown in the diagram, the new air being admitted through the valve V.

The cooling on free expansion, we have seen, is very small at ordinary temperatures—amounting for air to only a tenth of a degree Fahrenheit per atmospheric difference in pressure. But it increases as the temperature falls, so that the process goes faster and faster, the first hundred degrees being the hardest. The early Linde apparatus was excessively slow in getting under way. In the first run made in 1895, fifteen hours of continuous operation were required before the first drops of liquid air appeared. In order to shorten the time of the preliminary cooling, the air is now first cooled by a mixture of ice and salt or other freezing mixture. The regenerator has also been made more efficient by using three concentric pipes instead of two. The two compressors required have been combined into one double cylinder machine, which continuously pumps in a sufficient quantity of new air. With these and other improvements, a modern machine, compressing to 3000 pounds, will cool down in less than two hours, and thereafter produce a pint of liquid per horsepower per hour.

Liquid air has a temperature of −320 degrees Fahrenheit. It must therefore be kept in a vacuum flask. If exposed to the surrounding temperature, it quickly boils off. A favorite experiment is to place a kettle of liquid air on a block of ice. Clouds of steam pour from the spout. What we see, however, is not the air boiling off, but the surrounding moisture condensed by the stream of cold air. The liquid will boil more rapidly, though less spectacularly, if not placed on the ice. The latter merely provides a moister atmosphere. Similarly a small amount of the liquid dropped on the table disappears at once in a cloud of steam.

If a piece of metal is dipped in liquid air, and then touched by the finger, a severe burn will result. But a few drops of the liquid can be taken on the hand without discomfort, provided it is not allowed to *sit*. The reason is that the hand is in comparison a hot stove, and the liquid air assumes the spheroidal state. It rests on a cushion of its own vapor, a poor conductor, and does not really touch the hand.

Solid substances undergo curious changes at this low temperature. A soft rubber tube dipped in the liquid becomes hard and brittle and breaks into many pieces when hit on the table. A rose becomes like glass, and can be similarly shattered. A coil of lead wire becomes strong and elastic like spring brass, and will support a considerable load. But as the lead warms up, it loses it springiness, and is finally pulled out nearly straight by the weight. Mercury poured into the liquid is quickly frozen, and if a stick of wood is frozen into the mercury, a hammer can be made with which nails may be driven. If a piece of frozen mercury is dipped into a jar of water, Tyndall tells us[2] "it liquefies and showers down through the water; but every fillet of mercury freezes the water with which it comes in contact, and thus around each fillet is formed a tube of ice, through which the liquid metal is seen descending." Thus the water melts the mercury, and the mercury freezes the water.

In 1900, Dewar succeeded, by means of the apparatus of Linde reduced to laboratory size, in liquefying hydrogen. By placing a small dish of the liquid under an air pump, and exhausting rapidly, the hydrogen was further cooled by its own evaporation, and finally froze.

Hydrogen boils at − 423 degrees, and freezes at − 434, or at 37 and 26 degrees respectively above the absolute zero. The whole range of its solid and liquid states is thus confined to narrow limits. The liquid is colorless and transparent. It is extraordinarily light, being only one-fourteenth as heavy as water. The only solid that will float on its surface is pith. The liquid contracts on solidifying, the density of the solid being one-tenth that of water—less than half that of cork.

At these extremely low temperatures, some very remarkable experiments can be performed. If the cotton wool, which is used to stop the mouth of the vacuum flask containing the liquid hydrogen, is removed, a miniature snow storm occurs, formed by the *freezing* of the air which comes in contact with the intensely cold vapor that rises from the liquid. This snow falls into the vessel, sinks through the liquid, and collects at the bottom.

2 *Heat as a Mode of Motion*, p. 199.

If one end of a completely sealed tube filled with ordinary air is dipped into the liquid, the contained air soon freezes solid at the bottom of the tube, producing in the remainder an almost perfect vacuum. The only constituent of the atmosphere remaining unfrozen under these conditions is helium.

Temperatures below that of liquid hydrogen can not be produced by the Linde method alone. The latter must be supplemented by every other known means of producing cold—compression, sudden expansion, rapid evaporation stimulated by reduced pressure, etc. The procedure is first to produce liquid air and use that as a cooling agent, then liquid hydrogen for further cooling, and finally to apply any other methods available. In this way, Kamerlingh Onnes at Leyden succeeded in 1908 in liquefying helium at − 452 degrees, only 8 degrees above the absolute zero. For eighteen years he endeavored to solidify helium, and although he got the temperature down to 2.7 degrees absolute, the substance remained liquid. In the very year of his death, 1926, the feat was accomplished by his successor at the Leyden laboratory, W. H. Keesom. The latter produced the cold by rapidly evaporating the liquid helium in a vacuum, and then applied a pressure of 2100 pounds per square inch. The helium then froze at 7.6 degrees above the absolute zero. Later he succeeded in freezing it at lower pressures and temperatures, finally at 400 pounds pressure and 2 degrees above absolute zero. The freezing point of helium thus descends as the pressure is diminished, a behavior contrary to that of water. Possibly it cannot be frozen at all at atmospheric pressure. Thus the last of the "permanent" gases was conquered. Since the critical temperature of helium is − 450 degrees, and the absolute zero is − 460 degrees, the whole solid and liquid life of this remarkable element is lived within ten degrees of the absolute zero.

BIBLIOGRAPHY

Only the more important general works are here listed. Other references given in the footnotes are not here repeated.

ALTHOUSE, A. D., and TURNQUIST, Carl H. *Modern Electric and Gas Refrigeration* (Goodheart-Wilcox Company, 1933). ,

BACON, Francis. *Novum Organum* (1620).

CARNOT, N. L. Sadi. *Reflections on the Motive Power of Fire,* translated by R. H. Thurston (Dover, 1960).

CHALKLEY, A. P. *Diesel Engines* (D. Van Nostrand Company, 1915).

CLAUSIUS, Rudolph. *Die mechanische Wärmetheorie* (Third edition, Braunschweig, 1887).

DÜHRING, Eugen. *Robert Mayer, der Galilei des neunzehnten Jahrhunderts* (Chemnitz, 1880).

EWING, J. A. *The Mechanical Production of Cold* (Cambridge, England, 1921).

HELM, Georg. *Die Energetik* (Leipzig, 1898).

HELMHOLTZ, Hermann von. *Ueber die Erhaltung der Kraft* (Berlin, 1847. Reprinted in Ostwald, *Klassiker der exakten Wissenschaften,* No. 1, Leipzig, 1902).

———*Vorträge und Reden* (Braunschweig, 1865).

HEYL, Paul R. *Fundamental Concepts of Physics* (Williams and Wilkins Company, 1926).

HULL, H. B. *Household Refrigeration* (Nickerson and Collins, 1924).

JOULE, James Prescott. *The Scientific Papers of* . . . (London, 1884).

LENARD, Philip. *Great Men of Science* (The Macmillan Company, 1933). Chapters on and portraits of Carnot, Mayer, Joule, Helmholtz, Clausius, Kelvin, Papin, and Watt, among others.

LEWIS, Gilbert Newton. *The Anatomy of Science* (Yale University Press, 1926). See Chapter VI: "Probability and Entropy."

MACH, Ernst. *Principien der Wärmelehre* (Leipzig, 1896).

———*Die Geschichte und die Wurzel des Satzes von der Erhaltung der Arbeit* (Prague, 1872). English translation by Philip Jourdain: *History and Root of the Principle of the Conservation of Energy* (Open Court Publishing Company).

———*Popular Scientific Lectures*, translated by T. J. McCormack (Open Court Publishing Company).

MAXWELL, James Clerk. *Theory of Heat* (New Impression, Longmans, Green, and Company, 1908).

MAYER, Julius Robert. *Bemerkungen über die Kräfte der unbelebten Natur* (Liebig, *Annalen der Chemie und Pharmacie,* 1842).

———*Die organische Bewegung in Zusammenhang mit dem Stoffwechsel* (Heilbronn, 1845).

———*Beiträge zur Dynamik des Himmels* (Heilbronn, 1848).

———*Das mechanische Aequivalent der Wärme* (Heilbronn, 1851).

PLANCK, Max. *The Theory of Heat Radiation,* translated by Morton Masius (Dover, 1959).

REEVE, Sydney A. *The Thermodynamics of Heat-Engines* (The Macmillan Company, 1903).

RIPPER, William. *Heat Engines* (Longmans, Green, and Company, 1909).

STODOLA, A. *Steam and Gas Turbines* (Peter Smith).

SUPLEE, Henry Harrison. *The Gas Turbine* (J. B. Lippincott Company, 1910).

TAIT, P. G. *Sketch of Thermodynamics* (Edinburgh, 1868).

THOMAS, Carl C. *Steam Turbines* (John Wiley and Sons, 1906).

THOMPSON, Sylvanus P. *Life of Kelvin* (Macmillan and Company, 1910).

THOMSON, William (Lord Kelvin). *Mathematical and Physical Papers by* ... (Cambridge, England, 1882).

THURSTON, Robert H. *A History of the Growth of the Steam-Engine* (D. Appleton and Company, 1901).

TYNDALL, John. *Heat as a Mode of Motion* (D. Appleton and Company, 1863).

——*Fragments of Science* (Sixth edition, D. Appleton and Company, 1897), Volume II.

YOUMANS, Edward L. *The Correlation and Conservation of Forces: A Series of Expositions by Grove, Helmholtz, Mayer, Faraday, Liebig, Carpenter* (D. Appleton and Company, 1865).

WEYRAUCH, J. *Robert Mayer, kleinere Schriften und Briefe* (Stuttgart, 1893).

WOOD, Alexander. *Joule and the Study of Energy* (London, 1925).

INDEX

Absolute temperature scale, 107, 158, 160

Adiabatic, *defined*, 43; entropy constant, 162; is plumb-line, 166

Air cooler, 191

Alchemist, 142

Alcohol, as working substance, 54

Animism, 59, 141

Anticipators and discoverers distinguished, 123

Archimedes, lever, 124

Areas, are heat on temperature-entropy diagram, 165; measurement of, 23; are work on indicator card, 22

Assumptions, inescapable, 124

Atoms, 59, 123; denied, 99

Automata, 3

Automatic universe, 144

Bacon, Francis (1561-1626), cold as a mode of motion, 67; conspicuous instances, 66; heat as motion of parts of tolerable size, 66; induction, 65, 67; instances, 65; less hasty generalization, 67; method, 65

Bacon's hypothesis, converse of, 69; tested, 68

Bernoulli, Daniel (1700-1782), gas pressure, 73

Black, Joseph (1728-1799), latent heat, 9, 29

Boiling points and pressure, 29

Boltzmann, Ludwig (1844-1906), 168

Boulton and Watt, 13

Boyle, Robert (1629-1691), heat as a molecular motion, 72

Boyle's law, 46, 160

Branca, impulse turbine, 3

British Association electrical units, 106

British thermal unit, 28

Bucket chain analogy, 166

Caloric, 27

Caloric theory, death blow by Joule, 78, 130; doubted by Carnot, 27, 40; fric-

tional heat, 75; heat of compression, 85; heat content, 75, 77; inadequacies of, 115; in Mayer's time, 99; pulverization, 76; is quantitative, 74; specific heat at constant pressure, 87; specific heat of gas varies with volume, 40; specific heat of vacuum, 85

Calorie, *defined*, 28

Calorimeter and calorimetry, 28

Capacity, of Carnot gas cycle small, 173; of engine, *defined*, 11; sought by practical men, 25

Carnot, Nicolas Léonard Sadi (1796-1832), anticipates energy theory, 134; his book, 25; caloric theory doubted, 27, 40; compound expansion recommended, 16; efficiency often secondary, 56; experiments suggested, 135; fall of heat, 31, 74; heat is molecular motion, 134; internal combustion recommended, 56; inverse operations, 32, 34; latent heat, 30; loss between fire and boiler, 55; maximum motive power, conditions for, 35, 36; mechanical equivalent calculated, 135; motive power diminishes with temperature height of source (Carnot function), 40; motive power possible wherever temperature difference exists, 31, 32, 37; notebook, 57, 134; operations of perfect steam-engine, 33ff; perfect gas cycle, 42; perpetual motion excluded, 35; pre-compression recommended, 56; steam cycle, 42; volume changes required for motive power, 32; waterfall analogy, 36, 40, 74; working substance, 27, 32

Carnot cycle, composed of adiabatics and isothermals, 43; efficiency of, 154; fundamental equation of, 153, 161; motive power independent of working substance, depends only on temperatures, 35, 36; for perfect gas, 46, 151, 173; possible with any working substance, 43; is reversible, 42; small capacity for gas, 173; steam, 42, 44